Zoological Illustrations,

OR

ORIGINAL FIGURES AND DESCRIPTIONS

OF

NEW, RARE, OR INTERESTING

ANIMALS,

SELECTED CHIEFLY FROM THE CLASSES OF

Ornithology, Entomology, and Conchology,

AND ARRANGED ON THE PRINCIPLES OF
CUVIER AND OTHER MODERN ZOOLOGISTS.

BY

WILLIAM SWAINSON, F.R.S., F.L.S.,

MEMBER OF THE WERNERIAN SOCIETY OF EDINBURGH, ETC.

VOL. I.

1820

Zoological Illustrations. Vol I
by William Swainson
© 2021 la .rydy. ckupra
ISBN 978-1-716-19 429-0

Table of Contents

PREFACE.

The termination of the first volume of the Zoological Illustrations is accomplished, and its contents will not only enable our readers to discern the nature of the work, but likewise to form a judgement, on that degree of improvement which we have introduced since its first publication, and they may safely rely on the continuation being in no respect inferior.

In commencing a work of this nature, we have had two principal objects in view: the diffusion of original observations, which, while they might further the ends of science, would also be interesting to the general reader; and that of discouraging the publication of distorted figures copied from old authors, by accustoming the public eye to original designs and correct representations of natural objects. How far we may have succeeded in this latter object, remains to be judged by others; we are however satisfied with having made the attempt, and we hope that abler pencils than our own, may engage in the prosecution of this most desirable object; for it is only by the publication of original matter, that a check can be given to the increasing number of compilations and multiplied copies of "ill-shaped" figures, by which error is perpetuated, and science retarded.

The only original work that has appeared in this country similar to our own, is the Zoological Miscellany by Dr. Leach, which, as it was discontinued after the third volume, it may be presumed was unsuccessful: although little can be said of many of the figures in the early volumes, those in the latter are much to be praised, and the whole are original; the descriptions also abound with details highly interesting to the scientific world, for which indeed the learned author principally intended it; nevertheless it is a question, whether science in the end would not have been equally, and perhaps more advanced, had this work been more adapted to general readers. Instruction in these days of refinement must be made easy, palatable, and enticing; the eye must be pleased, while the understanding is improved, and Wisdom in her simple dignified garb will often be deserted for Ignorance, decked out in the glittering trappings of Folly.

The *Naturalist's Miscellany* conducted by Dr. Shaw, in its miscellaneous nature also resembled the present work, and reached to the extent of twenty-four volumes. What an invaluable fund of information these might have contained had their contents been original! Unfortunately, however, the exceptions are so few, that the whole may be termed a loose compilation, the descriptions being mostly given in as few words as possible, and the figures not only copied from wretched representations found in old authors, but often coloured from their descriptions only! It is indeed lamentable that the Author, whose talents and abilities were unquestionable, should have exerted them so little, and thus have descended to the rank of a voluminous compiler, for little better can be said of the General Zoology, begun and continued under his name: little original matter can there be found, excepting in the latter volumes, yet even in these no notice whatever is taken of the immense number of new species discovered in Africa by Le Vaillant, and long ago published in the *Oiseaux d'Afrique*: the engravings also are in like manner copied from old prints, enlarged or diminished as occasion offers, without even a regard to the selection of the best. It may be as well to observe in this place, that a great number of generic distinctions have been made in the two last volumes; which, as they have not been followed by any of the great and acknowledged Zoologists on the Continent, and appear to us in many instances trivial and unnecessary, will not be adopted in this work.

It will be unnecessary to point out with regard to the scientific arrangement, that we have avowedly adopted the principles of the modern classification; which the strict followers of Linnæus (in this country alone) have so long, but so ineffectually opposed. The first has been designated as the natural, and the other the artificial system; and, without entering into a critical disquisition on these definitions, it will be sufficient to observe, that by the Artificial System we bend nature to conform to certain arbitrary principles, which we lay down and to which we insist all her productions known and unknown will conform; while in the Natural method, we endeavour by tracing her modifications, to adapt our system to that which appears to regulate her operations. In the one we give laws, in the other receive them; by the first we are taught to believe that the highest attainment of the science, is that of ascertaining the name of an object in our Museum, or of giving a new one; we record it in our favourite system as a grammarian enters a new word in his dictionary, and there the matter terminates. Where the artificial system ends, the natural begins; for we then proceed to the investigation of affinities founded on anatomical construction, economy, and geographic distribution; our attention ceases to be confined to individuals, and extends to large groups; general facts enable us to draw general conclusions, till the mind begins faintly to discern a vast and mighty plan, by which the zones of the earth are peopled by their own respective races of animate beings; blending their confines unto each other with divine harmony, beauty, and usefulness.

That these inquiries and results have had a most wonderful effect on the natural sciences of late years, is abundantly evident. Geology, a subject hardly thought of in this country a few years ago, is now found to be a science of the first importance; with this, however, Conchology is so intimately connected, that without a certain knowledge of it, the geologist is frequently unable to prosecute inquiries of the most interesting nature; and there is little doubt but that Botany has been raised to the rank it now so justly holds, solely because its natural system has been more generally studied and advocated in this country, than that of any other branch of Natural History. In this science at least, we possess a superiority which our continental neighbours cannot dispute; and the name of Brown will be enrolled in the brightest page of our philosophic inquirers.

That the prejudiced adherence to the strict Linnæan system, has been the primary cause why Zoology has been more neglected with us than on the Continent, will admit of little doubt; for by shutting the door to all further improvement, it has impressed the generality of our countrymen with an idea, that the highest object of the Naturalist was to label the contents of a museum, and to arrange stuffed animals, like quaint patterns of old china, in glass cases: to thinking minds no less than to the vulgar, this idea has produced a feeling of contempt and ridicule, and very few of those qualified by nature for accurate investigation and philosophic reasoning, have been induced to make the science a study; and thus from such an unfortunate prejudice, to use the words of a powerful writer of the present day, "some future historian of the progress of human knowledge, will have to state that England, till within the few last years, stood still at the bottom of the steps where Linnæus had left her; while her neighbours were advancing rapidly towards the entrance of the temple[1]."

Finally—Linnæus to a comprehensive genius united indefatigable industry; yet he could not see and study those innumerable productions that have been discovered since his death: in proportion as our knowledge of objects increases, so must our systems change, until the natural one is fully developed; and the question simply comes to this, Whether the Linnæan method should be upheld as a solitary exception to the mutability of human wisdom.

The sun of truth must however finally prevail, and there is every reason to think it has already broke, and will gradually disperse these mists of prejudice. It is however much to be regretted, that our public institutions are wholly inadequate to facilitate not only the advancement of students, but the researches of those who are already engaged in prosecuting their inquiries: in Scotland alone are founded any Professorships of Natural History, and the establishment of our National Museum (in this branch only) is confessedly difficult: materials for study are more necessary in this science than any other; yet the public Institutions and libraries of the metropolis, "rich and rare" in every other department of knowledge, in most instances are deficient in this of the most elementary books; setting aside those of illustration, which, from being unavoidably expensive and within the reach of few purchasers, are more particularly adapted for such general repositories of learning. The protracted ill-health of its noble possessor, was the cause no doubt of the Banksian magnificent library being left deficient in several of the latest continental works; and that of the British Museum I have reason to think is still more defective. To the honour however of the keepers of the Bodleian and Radcliffe Libraries, it should be mentioned, that no pains or expense have been spared to render them as perfect in this branch as possible; and we have been told that the latter particularly is the most magnificent in the kingdom.

We shall now as briefly as possible advert to the contents of this volume.

In the Ornithological department the systems of Cuvier and Temminck have mostly superseded all others: as a whole, we give a decided preference to the latter, as being more natural, though it may be doubted if the generic distinctions are not too few, while those of Cuvier are

---

1   *Horæ Entomologicæ*, by W. S. MacLeay, Esq. M.A. of Trinity College, Cambridge. London, 1819. A work which for acuteness of reasoning and profound research, has never been equalled either in this, or perhaps in any other country. dd

too many: both however can be considered only as sketches, subject to improvement—as natural affinities are more studied.

Regarding that part of our work which relates to Entomology, we have given a decided preference to the Lepidoptera, for the simple reason that this order has received less attention from all writers, concerning their real characters and affinities, than any other; indeed they have been most unaccountably neglected even by Latreille, the great founder of the modern school: we have therefore thought it necessary to propose in this department many new genera, and only have to regret that their definitions could not be made more perfect without the destruction of the specimens, frequently not our own, and which therefore was unattainable: a more extended knowledge of the natural affinities existing in this tribe, will alone confirm or annul the propriety of these distinctions.

In *Conchology* many of the genera long established on the Continent, but new to our own collectors, have been characterized and illustrated, as well as specific distinctions defined between shells hitherto considered as varieties; and here it must be observed that so much latitude has been given to the meaning of the term *variety*, that in its general acceptation its definition becomes impossible: our own idea of its true meaning is, a shell possessing one or more characters which are changeable and uncertain, and which consequently will not serve as indications by which it may infallibly be distinguished from all others; *variety* depends on local circumstances, and affects the size, colour, and greater or less development of the same modification of structure; a *species* is permanent, its structure always the same though more or less developed, producing and perpetuating its kind, and depending on formation, discernible in youth, and matured in age: we cannot therefore comprehend the contradictory term of *permanent varieties* in a state of nature (though such occur in domesticated animals), which some authors have used, and which has led to, in many instances, the most erroneous conclusions.

It is lamentable to see the opposition which is still made by our own writers against all the modern improvements; yet although Linnæan Introductions to Conchology are constantly issuing from the press, the desire of being acquainted with a more natural and intelligible classification has already appeared; and as we are frequently questioned on the subject, we cannot in this place do better than refer the young student to the valuable article on Conchology contained in the late supplement to the *Encyclopedia Britannica*, the perusal of which will convey more solid information, and less perplexity, than all the Introductions our booksellers can boast of.

With respect to the quotations or synonyms, it should be observed that we have in most instances limited them only to original works, all doubtful ones have been rejected, and such only given as have been actually consulted; indeed to this latter cause must be attributed the occasional omission of some, existing in books we had not the immediate power of consulting; our own library is not small; but the difficulty and expense of procuring all the new continental publications, and the impossibility of meeting with them at our public libraries[2], may have sometimes led us into error, and unintentionally to have passed over the discoveries of others.

With the few additions contained in the Appendix we shall now conclude; trusting that in the remarks drawn from us by the present state of the science in this country, our zeal for truth will not give us an appearance of want of candour or of vanity. The truth of our remarks on the labours of others, every one at all acquainted with the subject can inquire into, and either acknowledge or disprove: we neither deprecate nor despise criticism: an author who presumes to instruct others, should have his pretensions publicly canvassed, his merits admitted, or his deficiency exposed; no one is more sensible than we are that our own pretensions chiefly consist in having set an example for others more able to follow: and if we have in any way advocated the cause of truth and science, our object will be attained, and we shall then gladly retire in the shade.

London, Sept. 15, 1821.

---

2   It is truly grievous in those which are privileged to possess themselves of the works of their countrymen, however expensive, at free cost, and thus to inflict a ruinous fine on authors. Thus—National Institutions, founded for the encouragement of learning, are made to oppress and impoverish its followers.

LIST OF BOOKS REFERRED TO.

Bruguire, Encycl. Meth. Histoire Naturelle des Vers, par M. Bruguire, 1 vol. 4to. and 4 vols. of Plates, forming part 10, 19, 21, 23, of the "Encyclopedie Méthodique." Paris, 1789-1792.

Bloch. Histoire Naturelle des Poissons, en 6 parties, 8vo. Berlin, 1796.

Cramer. Papillons Exotiques, 4 vols. 4to. Amsterdam, 1779-1782.

Cuvier. Le Règne Animal, 4 vols. 8vo. Paris, 1817.

Chemnitz, Martini. Neus Systematisches Conchylein Cabinet, 11 vols. Nurnburg, 1781-1795.

Dill. A Descriptive Catalogue of Recent Shells. By F. W. Dillwyn, 2 vols. 8vo. London, 1817.

Edwards. A Natural History of uncommon Birds, &c.; and Gleanings of Natural History. By G. Edwards, 7 vols. 4to. 1763, &c.

Fab. Ent. Syst. Entomologia Systematica, emendata et aucta. J. C. Fabricius, 4 vols. 4to. Hafnia, 1792-1794.

Gen. Zool. General Zoology, commenced by Dr. Shaw, and continued by Mr. Stevens, 11 vols. 8vo. to 1819.

Gmelin Linn. Syst. Nat. C. Linné Systema Naturæ. Cura J. F. Gmelin. Lipsiæ, 1788-1793.

Godart in Encycl. Method. Encyclopedie Méthodique, t. 9. p. 1. 1819.

Gualtieri. Index Testarum Conchyliorum quæ adservantur in Musæo N. Gualtieri. Florentiæ, 1742.

Illiger. Prod. Systematis Mammalium et Avium, 8vo. Berolini, 1811.

Knorr. Les Delices des Yeux et de l'Esprit, 6 P., 4to. Nuremb. 1760, &c.

Klein Hist. Pisc. Historiæ Nat. Piscium promovendæ Missus, 6, 4to. Dantzic, 1740-49.

Linn. Syst. Nat. See Gmelin.

Linn. Trans. Transactions of the Linnean Society of London, 13 vols. 4to. 1791-1821.

Lister. M. Lister Historia Conchyliorum, folio. Oxonii, 1770.

Lamarck Syst. Hist. Nat. des Animaux sans Vertèbres. Par le Chevalier de Lamarck, 6 vols. 8vo. Paris, 1815-1819.

—— *Anal. Mus.* Annales du Museum d'Histoire Naturelle de Paris, 4to. 1802-1821.

Lath. Synop. Suppl. A General Synopsis of Birds. By Dr. J. Latham, 3 vols, and 2 Supplements, 4to. London, 1782, &c.

—— *Index Ornith.* Index Ornithologicus, 2 vols. 4to. London, 1790.

Martyn Univ. Conch. The Universal Conchologist. By T. Martyn, 4 vols. 4to. London, 1784, &c.

Martini. See Chemnitz.

Pennant. British Zoology. By T. Pennant, 4 vols. 8vo. London, 1812.

Risso Icth. Ichtyologie de Nice, 1 vol. 8vo. Paris, 1810.

Rumph. Thesâurium Imaginum Piscium, &c., folio. Hagæ, 1739.

Seba. Albertus Seba Rerum Naturalium Thesauri, 4 vols. folio. Amsterdam, 1734-1765.

Say. Description of the Land and Fresh-water Shells of the United States. By Thomas Say. Philadelphia, 1819.

Shaw in Gen. Zool. See General Zoology.

Temminck Pig. et Gall. Histoire Naturelle Générale des Pigeons et des Gallinaces. Par C. J. Temminck, 2 vols. 8vo. Amst. 1813.

—— *Manuel.* Manuel d'Ornithologie, 2d edit., 2 vols. 8vo. 1820.

Le Vaill. Hist. Nat. des Toucans et des Barbus, folio. Paris, 1806.

—— Hist. Nat. des Perroquets, 2 vols. folio. Paris, 1801.

White's Voyage. Journal of a Voyage to N. S. Wales, 4to. Lond. 1790.

1. *Psittacus Cayennensis.*
Cayenne gold-winged Parakeet.

Generic Character.

Bill short, thick, very strong, covered at the base by a cere; upper mandible sharply hooked; under mandible obtuse, curving upwards, and much shorter. Nostrils round, naked, nearly vertical. Feet scansorial.

Specific Character.

Green Parakeet, with the spurious wings golden-orange: outer quills blue at their base. Tail short, wedged. [Le Perruche à tache souci. *Le Vaill. pl.* 58, 59. *p.* 169]

We only thoroughly identified this beautiful little bird, by an inspection of the costly work of Le Vaillant on this family, in the Banksian Library: for the description of the orange-winged Parakeet of Dr. Latham is not applicable; and Dr. Shaw has persisted in the old error of considering this species a variety of the Toui Parakeet, although the question had been put at rest by the original description and sound reasons of Le Vaillant. There is a wide difference between naturalists who compile, and form their theories from books, and those who study nature, and think for themselves; and nothing will result from the first but mischief to the science, and perplexity to the student.

Our figure is from a specimen brought from Demerara by C. Edmonston, Esq.; another is in the possession of A. MacLeay, Esq. Though rare in our cabinets, M. Le Vaillant says it is common in Cayenne. He has given a beautiful figure of the female, which is entirely green.

Total length six inches. Plumage above entirely green, beneath paler and inclining to yellow; just under the lower mandible is a small snuff-coloured spot, and a very narrow line of the same in front just above the nostrils; the quills dark-green, the greater ones on their outside base are blue, with which the head is also tinged. The spurious wings are entirely of a rich and clear orange. Inner wing-covers green. Quills inside greenish-blue, except on each side the shafts, where there is a line of black. Tail short, cuneated, hardly projecting an inch beyond the wings, both above and below green: the interior margin dirty-yellow, the feathers pointed. Bill and legs flesh-colour.

2. *Sitta frontalis*. Blue Nuthatch.

Generic Character.

Bill very straight, sharp-pointed, compressed; both mandibles equally inclining to the tip, which is entire, and resembles a compressed wedge. Nostrils basal, oval, open, covered externally with incumbent setaceous feathers. Feet with three toes forward and one backward; inner toe very small; outer toe connected to the middle at its base; hind toe lengthened, strong. Claws much compressed; anterior nearly equal, posterior largest. Tail short, of twelve nearly equal feathers. Generic Type *Sitta Europæa*.

Specific Character.

Nuthatch, above blue: line above the eye, front, and middle of the lateral tail-feathers black; beneath cinereous brown, ears lilac, chin white.

The present species is one of the many interesting birds collected in Java by my friend Dr. Horsfield: it was not, however, until I had described and engraved another specimen, sent to Sir J. Banks from Ceylon, that I discovered the species had already been included in the Doctor's account of the birds of Java, presented to the Linnæan Society, where he has described it under the name of *Orthorynchus frontalis*.

The specific name of its first describer is of course retained: with respect, however, to its generic situation, I must be allowed to dissent from considering it as a distinct genus, merely from the prolongation of the hinder toe being somewhat more developed than in *Sitta Europæa* and *Carolinensis*, both which birds are now before me, and which in themselves differ in the relative proportion of this part: thus in *S. Carolinensis* the hind toe and claw is two-tenths of an inch shorter than the leg; in *S. Europæa* it is one-tenth shorter; and in the present species it just exceeds that of the leg: in every other respect not the slightest difference I apprehend will be observed.

Total length five inches. Size of the European Nuthatch. Bill, from the angle of the mouth to the tip, eight lines; front of the head velvet-black, continued in a stripe of the same colour over the eye, and terminating above the ear feathers: the upper plumage is of a rich blue, more brilliant on the head, and paler on the front, and external margins of the quills. Spurious wings and lesser quills black margined with blue. Inner wing-covers deep black; the under plumage is a light-brown, changing to lilac on the ears and sides of the neck, and tinged with cinereous on the flanks and vent: the chin is white; tail even, the two middle feathers blue, the rest more or less black, having the external margins and tips blue. The outer quill of the wings is short, the second and third longest and equal, the fourth rather less; the hind toe with the claw, measures one inch in a straight line.

3. *Mitra zonata.*
Zoned Mitre.

Generic Character.

Shell unequally fusiform. Spire lengthened, attenuated. Outer lip simple, not toothed within. Columella plaited.

Specific Character.

Mitre, with the epidermis marbled with brownish-yellow; volutions at their base black; columella five-plaited. [*Marryat in Linn. Trans. vol.* xii. *pl.* 10. *fig.* 1. 2.]

This unique and beautiful Mitre has been already described by Captain Marryat in the Linnæan Transactions: the figures, however, are uncoloured, and give a very indifferent idea of the graceful symmetry of its form. My friend Dr. Leach, with his usual liberality, permitted me to draw the accompanying figure of it at the British Museum, where it is now deposited.

It appears to have been taken near Nice in the Mediterranean, adhering to a sounding-line, in very deep water; a very singular locality, since nearly all the Mitres have generally been supposed to inhabit the tropical seas, or at least far from the coasts of Europe. It should, however, be remarked, that *Cypræa lurida*, an Asiatic shell, I have found on the shores of Greece: and C. Ulysses, in his travels in the kingdom of Naples, enumerates several shells as inhabiting the warm shores of the Tarentine Bay, which are generally known only as natives of the Red Sea and Indian Ocean. These facts, with many others, prove the physical distribution of *Molluscæ* to be less decidedly marked than almost any other class of animals.

This genus is included with that of *Voluta* by Linnæus and our own writers, although long ago justly separated by the continental zoologists.

4. *Bulimus melastomus.*
Blackmouthed Bulimus.

Generic Character.

Shell oval or oblong-oval. Spire elevated. Mouth entire, sub-oval. Column smooth, simple. Exterior lip thick, reflected. Interior lip beyond the middle inflected, and hollowed beneath. Operculum none.

Specific Character.

Shell oblong-ovate, white, marbled with cinereous. Spiral whorls longitudinally plaited. Outer lip flattened; aperture black.

The genus *Bulimus* was long ago formed by Scopoli out of the heterogeneous mixture of shells thrown together in the Linnæan genus *Helix*[3], &c.: it comprehends some of the larger and most beautiful of the exotic land shells, among which the present species will stand conspicuous both in beauty of colouring and excessive rarity. While travelling among the forests of Brazil, in the province of Bahia, I found the shell here figured one morning on the leaves of a *Solanum.* I not only searched myself, but promised as a reward to any of my Indians who would bring me another, a two-bladed Birmingham knife!—the greatest temptation they could have!—but in vain; for I never saw another before or since.

There are many peculiar characters presented in this species independent of its colour: the spiral whorls are strongly plaited longitudinally about half their length, and marked very slightly (but sufficiently distinct) with several oblique indented striæ; the principal whorl has a row of indented and unequal sulcations near the suture, and a slight appearance of elevation along the white transverse band; the outer lip is thick, broad, and flattened beneath; but the margin is reflected back, and forms a prominent rim on the upper surface. The shell, when viewed closely, appears rough with minute scale-like elevations, very much resembling shagreen.

---

3   Were it necessary at this time of day to point out the unnatural separation of shells intimately connected with each other, which the Linnæan arrangement presents, it would be sufficient to observe, that the genus *Bulimus* is formed of shells scattered in the old genera of *Turbo, Helix,* and *Bulla*: thus we see in Mr. Dillwyn's Catalogue, the large pink-mouthed African land-snail put in the same genus with our English *Bulla lignaria*, and *aperta*; the one inhabiting the depths of forests, and the others the depths of ocean!

5. *Colias Statira.*

Generic Character.

Palpi short, curved, compressed on the tongue, thickly covered with scales. Articulations three; the first very long, curved at the base, erect beyond; the second erect, short; the third minute, inclining forward; the tip naked, obtuse. Antennæ short, cylindric, gradually thickening to their tip, which is naked and abruptly truncate. Anterior wings trigonal. Abdomen of the male with the last joint pointed, and a slender incurved hook beneath; the valves large, attenuated and hooked. Generic Type *Colias Ebule.*

Specific Character.

Wings diluted yellow or fulvous; anterior with a black border and central dot, which beneath is ferrugineous; posterior beneath, each with two unequal snowy spots; palpi lengthened.—*Female.* [Papilio Statira. *Cramer, pl. cxx. fig.* C. D.]

T he present insect is selected to illustrate a very elegant family of Butterflies, whose predominant tints are composed of orange, yellow, and white, variously blended and disposed in a greater or less degree throughout all the species. The generic characters above given will distinguish them as peculiar to the tropics, and principally those of South America; one or two species only being found in Africa, and five or six inhabiting India.

I have no doubt this is the *Pap. Statira* of Cramer; it is found only in Brazil, and has been erroneously considered by Godart and Latreille as a variety of *C. Jugurthina*, an Indian insect, and which in fact is not in itself a species, being no other than the female of *C. Alcmeone*, as an attentive examination of a vast number of both, collected in Java by Dr. Horsfield, enabled me to ascertain.

The extraordinary prolongation of the last joint of the palpi, and the white borderless spots beneath, which are never silvered, will distinguish this species through all the variations; in the ground colour of its wings, which in no two specimens are exactly alike, and one before me is nearly white; the lesser snowy dot is sometimes very obscure, and often wanting; but the prolongation of the palpi is even expressed in Cramer's figure above quoted.

I have examined about a dozen specimens, mostly captured by myself, and all have been females; and I strongly suspect future and more decided observations will prove *C. Evadne* to be the other sex: it has the palpi lengthened, though in a less degree; and the articulations of the antennæ in both insects will be found somewhat thickened at their termination when viewed under a magnifier, a peculiarity I have seen in no other species; and although I have examined near thirty specimens of *C. Evadne*, they have invariably proved to be males.

The palpi in this insect will be found at variance with the generic character now given; a striking proof that in a natural system no single part can be taken as an unerring criterion for generic distinction, without making it eventually an artificial one. The *Colias Drya* of Fabricius has the same formation of palpi, but is a totally different insect.

6. *Colias Leachiana.*
Leachian Colias.

Generic Character.

See Plate 5.

Specific Character.

(Male) wings slightly rounded, entire, greenish white; anterior pair above orange, at their tips, margin, and central dot black: each pair beneath with a central ferrugineous spot. Female ——?

[C. Leachiana. *Godart in Encycl. Method.* vol. ix. p. 91.]

In size this insect is the largest of the genus yet discovered; it was first noticed by Godart, who has given it the name of my learned and valued friend, Dr. W. E. Leach, of the British Museum, whose talents are too well known to need any eulogium in this place.

It appears to inhabit both the northern and southern extremities of Brazil; for I have seen it in a box sent from Parà, and my specimens were captured in Minas Geraes by my friend Dr. Langsdorff. It is, however, a rare species; for I have only seen seven or eight specimens, and they were all males: the female, when found, will probably differ as remarkably as in most of this genus.

The opaque spot on the inferior wings above is very large; but the tuft of hair corresponding beneath the superior wings, is entirely wanting. It should be likewise observed, that although this insect in every outward respect resembles a genuine *Colias* (the type of which may be *C. Ebule*), it differs very materially in the terminal appendages of the abdomen; the last joint being the shortest, and scarcely pointed; and the hook, instead of being concealed beneath this segment, is exserted beyond it, and met by two others, one at the base of each lateral valve: these valves are also much shorter, ovate, and not attenuated, although ending in an incurved hook. In the present ignorance in which a true knowledge of the Lepidoptera is involved, it is impossible to say how far these dissimilarities may point out natural groups; it is therefore of the highest importance to the science such facts should be noticed.

7. *Carduelis cucullata.*
Hooded Seed-eater.

Generic Character.

Bill short, stout, very conic, without any curvature above; both mandibles nearly equal, the tip entire, straight and sharp; upper mandible convex above: lower one at the base of the margin with
an obtuse angle, the sides and under part convex. Generic Type *Fringilla Canaria.* Latham, &c.

Specific Character.

Orange: head, front of the neck, bar across the wing-covers, quills and tail black; greater quills at
their base obliquely barred with orange.

A richly coloured little bird, much smaller than our Goldfinch, and approaching very
near to the *Bouvreuil de Bourbon* of Buffon, from which, however, I think it quite distinct. The only one I have yet seen is in the possession of E. Falkner, Esq. of Fairfield near Liverpool, who received it with a few other rare birds from the Spanish Main.

Total length four inches. Bill blackish and very sharp. The whole head and forepart of the
neck is black. The plumage of the body is a fine reddish-orange, duller on the back and brightest beneath: wing-covers the same; the greater ones at their base black, which forms a bar:
the quills are also black, the greater ones having at their base an oblique bar of orange, and
some of the lesser ones slightly margined externally with white. Tail divaricated and black;
some of the lateral feathers faintly margined with orange. Spurious wings black. Legs and
claws brown.

The *Bouvreuil de Bourbon*, and the *B. du Cap de Bonne Esperance* of Buffon (*Pl. Enl. pl.* 204. *fig.*
1, 2.) appear to have been described as the different sexes of one bird (the Orange Grossbeak
of Latham) on mere conjecture. I think them quite distinct, inhabiting different countries, and
having all the appearance (in the figures) of being two male birds; for the females in this family seldom possess the rich colours of the male; and the figure of the last of these birds, has
not the slightest habit of a female.

The present genus was formed by Cuvier, (though but very slightly defined,) and includes the
common Goldfinch and Canary-bird.

8. *Merops urica.*
Javanese Bee-eater.

Generic Character.

Bill lengthened, smooth, slightly curved, terminating in a sharp point; the base triangular, the sides much compressed, the back carinated. Feet very short, gressorial. Wings pointed. Generic Type *Merops apiaster.* Linn., &c.

Specific Character.

Green, beneath paler. Head and neck above rufous; chin and throat sulphur; line under the eyes, and collar round the neck, black. Tail-covers and rump pale blue. Tail slightly forked.

[Merops urica. *Horsfield in Linn. Trans.*]

The true Bee-eaters are confined to the old world, principally inhabiting Africa and Asia; one species only, the European Bee-eater, being known with any degree of certainty to be found in Europe; and this is occasionally seen in England. They are all gregarious, feeding on the wing, and in general migratory.

Most unwillingly I have again in this instance anticipated my friend Dr. Horsfield in describing this bird, which he found in Java, and which I engraved after one sent from Ceylon, without knowing it had also fallen under his observation.

The figure is less than the natural size, which is nearly that of our European species. Bill an inch and a half long from the gape, and black. Nostrils small, basal, round, not sulcated, partially defended by incumbent hairs; at the angle of the mouth is a row of short, stiff bristles; a black line commences from the nostrils, passes under the eye, and terminates with the ears. The upper part of the head, neck, and between the wings, rufous. The rump and upper tail-covers pale blue: the chin and throat sulphur tinged with rufous, where an irregular and narrow collar of black crosses the neck. The remaining under parts yellowish-green. Wings and quills fulvous green, the latter tipt with black, and all the inner shafts more or less rufous: the second quill longest, and the lesser quills and tail-feathers notched at their tips. Tail green, slightly forked; the tips and under side dusky-black, and three inches and a half long. Wings, when closed, four inches one line in length. Vent blueish-white.

The females in this genus may generally be distinguished by the two middle tail-feathers being but slightly or not at all elongated.

9. *Helix auriculata.*
Eared Helix.

Generic Character.

Shell orbicular or globose. Spire depressed, or but slightly elevated. Aperture entire. Outer lip margined. Operculum, none.

Specific Character.

Shell much depressed, marbled and doubly-banded with ferrugineous. Umbilicus large, deep. Aperture ear-shaped. Outer lip thickened, reflected, with a gibbous obsolete tooth within.

A shell no less remarkable for its form than its extreme rarity. The mouth bears a most striking resemblance to the human ear; and the only specimen known in this country is the one here figured, from the cabinet of Ch. Dubois, Esq., who obligingly favoured me with it for examination; neither does the exquisite work on the Land Shells, by M. de Ferrusac, now publishing at Paris, contain this species among the numerous matchless figures already given of this family.

In the present uncertainty respecting the natural groups of the genus *Helix*, as left by Lamarck, I have preferred for the present following the example of Cuvier and de Ferrusac, in placing it with that family, in preference to adopting the ill-defined and palpably artificial distribution of them by D. de Montfort, or of forming a new genus for its reception.

The variegations in its colouring are better seen in the figures than described. The whole shell is slightly marked with obsolete longitudinal striæ; the umbilicus is very deep, and the tooth does not extend externally beyond the margin of the lip.

10. *Strombus.*

Generic Character.

Shell ventricose; base with a short canal, which is either emarginate or truncate; external lip dilated into a simple wing, notched at the base, and prominent above. Animal marine, carnivorous; body spiral, with a compressed foot at the inferior base of the neck. Generic Type *S. pugilis.* Linn.

Little Strombus—central figures.

Shell with nodulous plaits; the spire finely striated; inner lip thickened and reflected, and obtusely pointed above. Outer lip smooth within, deeply lobed above, attached to the second spiral volution. [*Lister* 859. 15. *Chemnitz. tab.* 156. *fig.* 1491, 1492. *Rumph. tab.* 36, P. *Gualtieri, tab.* 32, G.] [Strombus marginatus. *Dillwyn's Cat.* p. 665. no. 18.]

A pretty and diminutive species, scarcely ever more than one inch three lines long. The spire long in proportion, and occupying half an inch: when in perfection the colour is a deep chesnut, minutely broken into finely serrated darker lines, with one, two, or three interrupted bands of white on the body whorl, the spire, and margin of the outer lip paler; there are two or three nodules above; and the spiral volutions have the carinated row of tubercles usual in the *Strombi*, and are besides finely striated transversely. The base of the shell is more deeply and distinctly striated; both the lips are much thickened, tumid, white, and highly polished; terminating above in obtuse points on the second spiral whorl, leaving a narrow ascending channel between; the inside of the aperture is a fine yellow. Inhabits the Indian seas, but is not common.

Variable Strombus—upper figure.

Shell with nodulous plaits, the spire not striated. Inner lip simple. Outer lip reflected, smooth within, and slightly lobed above.

S hell two inches and a quarter long, the spire occupying little more than half an inch. The ground colour generally is white with numerous undulated short lines of a darker colour, sometimes crossed by four or five obsolete whitish bands: it approaches very near *S. minimus*, but is easily distinguished by being in general much larger, by having the inner lip not at all thickened above, the outer lip very slightly lobed, and only advancing on the first volution of the spire: it varies, however, amazingly in colour. There is a small variety, having a brown spot beneath, from India; and others (labelled from the So. Seas) in the Banksian collection, also small, are purplish-brown, with three or four well-defined bands of white: the aperture is always pure white.

11. *Drusilla Horsfieldii.*

Generic Character.

Antennæ moderate, the club lengthened, slender, cylindric. Palpi short, much compressed, obtuse, remote, not touching the tongue, covered equally on both sides with thickset hairs concealing the joints. Abdomen (in the male) 7-jointed, the first very long, the last simple, truncate, and entire above, without valves, and with an incurved hook beneath. Anterior wings (in the male) dilated at the posterior base, concealing a tuft of hair on the inferior wings.

(Obs. Wings very entire, sub-diaphanous. Hinder wings large, orbicular. Fore-legs spurious. Basal articulations of the antennæ thickened at the end.) Generic Type *Papilio Jairus.* Fabr.

Specific Character.

Wings, anterior, narrowed, the posterior and exterior margins equal, uniform brown; posterior cream-coloured, with a brown margin, one ocellate spot above, and two beneath.

This new and elegant insect was discovered by my worthy friend Dr. Horsfield (after whom I have named it) in the interior of Java: it forms a part of the extensive collections made there by this zealous naturalist for the East India Company, and which will make a most important addition to our present confined knowledge of the productions of that interesting island: indeed these collections exceed in extent, preservation, and value, any which have been brought to this country.

The general resemblance of this species with *Papilio Jairus* of Fabricius is so great, that it might pass on a cursory view as a mere variety, did not the form of the anterior wings at once point out the difference. In the present insect the posterior and exterior margins are of equal length, giving a narrow appearance to the wings, much resembling some of the true *Papilionidæ*; but in *P. Jairus*, these wings are much broader, more obtuse, and the length of the posterior margin much greater; other minor differences exist in the colouring and markings. My valued friend A. H. Haworth, Esq., F.L.S., &c. possesses a fine pair of the true *P. Jairus* in his rich and extensive collection; and the liberality with which it is constantly and freely opened to me, deserves my warmest thanks.

From these two species I have formed the present genus, the characters of which will distinguish it from that of *Hætera* (Fabr.), a scanty genus confined to South America; and this seems to occupy its place in India: both will come in the natural family of *Hipparchidæ*. Only one specimen (a male) exists in Dr. Horsfield's collections; I could not therefore dissect the mouth, &c. On the inner borders of the inferior wings is a lengthened tuft of fine hairs: the *anus* beneath has on each side an obtuse lengthened process, partially attached, and which appears to supply the place of the lateral valves. It is represented on a sprig of Gærtnera *racemosa*, which, although differing slightly from the figure of Dr. Roxburgh's Coromandel Plants, p. 19. t. 18, Dr. Horsfield considers as the same plant. The Javanese name is *Kakas*.

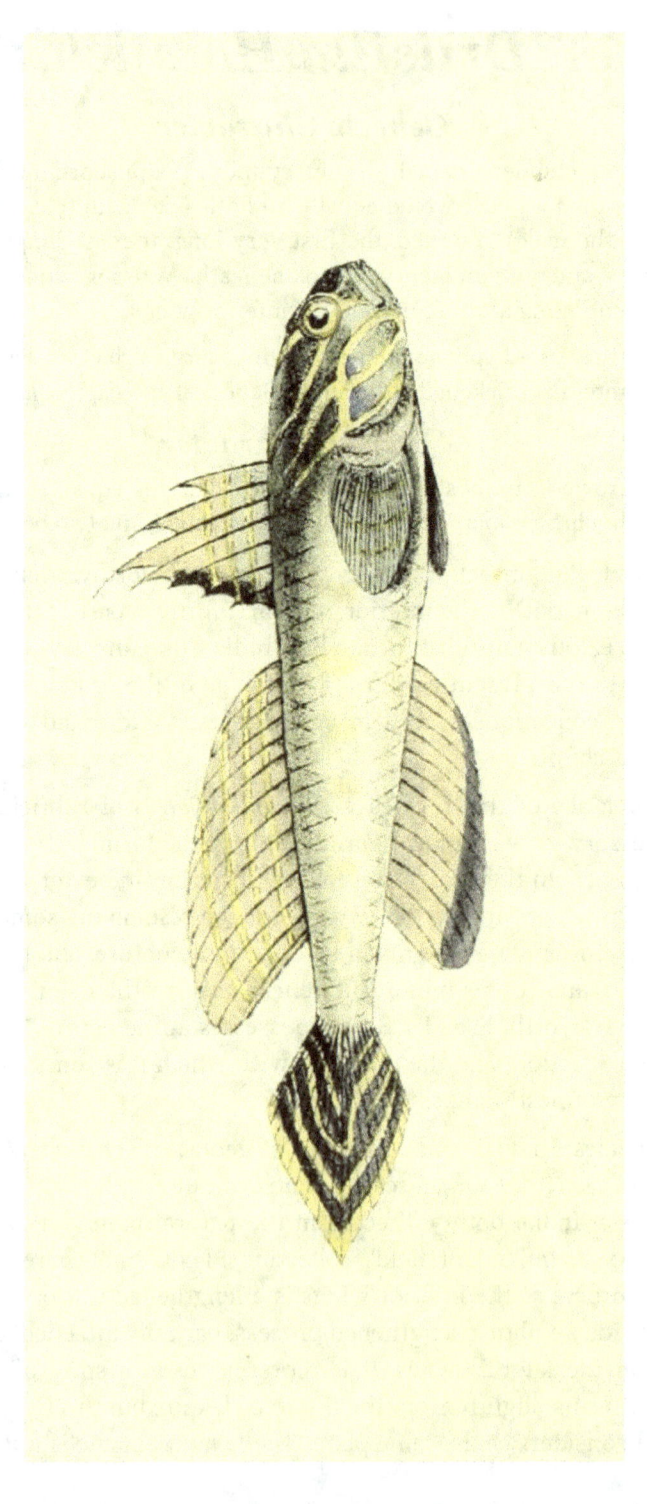

12. *Gobius Suerii.*
Suerian Goby.

Generic Character.

Head small. Eyes approximating. Dorsal fins two, the rays flexible. Ventral fins united into the form of a funnel. Gill aperture contracted, the membrane four-rayed. Generic Type *Gobius niger.* Pennant.

Specific Character.

Olivaceous Goby. Tail obscure purple. Head, gills, dorsal and caudal fins with yellow longitudinal lines. [Gobius Suerii. *Risso Icth. p.* 387. *pl.* 11. *fig.* 43.]

This beautiful little fish never exceeds the size here represented. It is not uncommon on the coast of Sicily in the spring months: it has likewise been discovered on the shores of Nice by Risso, who is its first describer: his figure, however, is so remarkably bad, that it would be impossible to recognise it but for his description: in fact, the fish is so delicate, that unless the fins are very carefully expanded in water their form and colouring will never be seen correctly. It has been named after M. C. A. Le Sueur (who accompanied Peron in the French circumnavigation), an able zoologist and most inimitable draftsman and engraver.

General colour pale olivaceous yellow, with a few obscure large spots along the body somewhat brighter. The head has four yellow oblique bands, between which, and behind the eye, is a bright blue spot. Ventral fins blackish. The first dorsal fin is trigonal, and has the third ray lengthened and longest; the three last rays tipt with deep-black; the second fin is lengthened, broadest at the end, the membrane greyish-white with narrow longitudinal lines of yellow. Caudal fin lanceolate, pointed, blackish-purple, margined and marked with yellow lines parallel with the borders. Pectoral fin ovately rounded, with faint yellow transverse bands. Anal fin resembling the second dorsal, greyish, with the margin dull-purple. Scales large, very deciduous; lateral line invisible; lower jaw longest; teeth minute. Body and fins semi-transparent. The first dorsal fin has seven rays; the second fifteen; anal fifteen; pectoral twelve; caudal seventeen; ventral fourteen.

15.

13. *Platyrhynchus Ceylonensis.*
Ceylonese Flat-bill.

Generic Character.

Bill short, straight, thin, very depressed, and nearly triangular; the upper mandible abruptly hooked at the tip, and notched; the margins folding over those of the under mandible, which is straight and shorter. Mouth and nostrils defended by long stiff bristles. Nostrils medial between the tip and gape of the bill. Tail mostly even, of twelve feathers. Legs and toes short, slender. Generic Types. Div. I. *Todus Platyrhynchos.* Gm. Div. II. *Muscicapa barbata.* Lath.

Specific Character.

Olivaceous Flat-bill, beneath yellow. Head and chin cinereous.

T he sober tints of this little bird accord more with those of Europe than of India, of which country however it is a native, having been sent from Ceylon to the British Museum: it is the only one I have yet seen, and appears hitherto undescribed.

The stiff bristles at the corner of the mouth are nearly the length of the bill, which is quite flattened: the tail is even, and the whole bird in every respect but colour closely resembles the bearded Flycatcher (*Musc. barbata* Lath.).

Cuvier and other modern zoologists have done much in distributing the Linnæan *Muscicapæ* into their natural families; but as we are acquainted with a great number from descriptions only, the arrangement is by no means perfect.

The generic characters now given of the genus *Platyrhynchos* (very slightly noticed by Vieillot) will be found perfectly applicable to the separate divisions here formed; the first comprising the *Todus Platyrhynchos* of Gmelin, and a few others having the bill larger and more dilated than the second division, which includes the present species, together with *M. barbata, cærulea, cuneata*, and no doubt many others. The construction of the bill in all these birds will be found precisely the same, though more or less developed in each division, and even in the species; it thus becomes impossible to draw the line of demarcation without refining too much on generic distinctions. Their bills, although so broad, are by no means stout; thus enabling them to prey with greater readiness on the *Lepidoptera* and other large winged insects with soft bodies; while the long stiff bristles at the base of the bill seem intended to confine the resistance their prey would otherwise make by their wings. The illustrious Cuvier has well observed, that the true Flycatchers have the bill longer, narrowed, less compressed, and the tip but slightly bent.

Pl. 14.

14. *Picus rubiginosus.*
Brown Woodpecker.

Generic Character.

Bill many-sided, straight, the tip resembling a compressed pointed wedge. Nostrils basal, oval, open, covered externally with narrow recumbent feathers. Tongue very long, retractile, the tip barbed. Tail-feathers ten, strong, rigid, acuminated; the two middle ones longest. Feet climbing. Generic Type *Picus viridis.*

Specific Character.

Above tawny rufous. Crown blackish; hind head crimson, beneath fulvous, with brown transverse bands.

The Woodpeckers form a most natural family of birds, and are dispersed in every part of the known world, excepting the Polar regions. Eight species inhabit Europe, five of which are found in our own country. The largest however of these, the Great Black Wood-pecker, is very rare; and even the others are less frequently seen than formerly, from the gradual diminution of our few remaining forests.

The present appears an undescribed species, and was sent from the Spanish Main to E. Falkner, Esq. of Fairfield. I have since seen the male, which, like many others of this genus, is distinguished by a patch of red below the eye.

Total length, eight inches and a half; bill one inch long, blackish; front and crown cinereous black; the hind head and nape crimson; a dusky whitish line (beginning at the nostrils) in-cludes the eye and ear-feathers; below this on each side blackish, with longitudinal whitish dots, which in the male is mixed near the bill with crimson; chin blackish, speckled with white. The general plumage above is uniform tawny rufous brown, becoming more olive on the rump. Under parts olivaceous yellow, crossed with numerous close bands of blackish brown. Quills with the inner web black, the margin pale yellow; shafts and outer web tawny; tail the same, the shafts and outer half black, excepting the last pair, which have yellowish shafts and dusky tips. Wings inside, pale orange. Legs and claws dusky green.

Two or three other individuals have since fallen under my observation: the male I saw at Mr. Leadbeater's, Animal Preserver, in Brewer-street, of whose liberality and integrity in every way, I can bear the most unqualified and cheerful testimony.

15. *Licinia Melite.*

Generic Character.

Antennæ slender, the club elongated, fusiform, and compressed. Palpi very short, hardly projecting beyond the head, compressed on the tongue, covered with scales and margined externally with long hairs, the last joint nearly naked and almost as long as the second joint. Body elongated, slender, in the male with 6 joints, the last entire. Valves generally elongated, attenuated, their tips acutely pointed. Anterior wings (in the male) narrow, obtusely attenuated; in the female broader, and obliquely rounded. Posterior (in the male) dilated, nearly as long as the anterior wings, the fore margin opaque; in the female shorter, and nearly orbicular. Generic Type *L. Melite.*

Specific Character.

Male. Wings yellow, anterior above black, with an oblique yellow band and transverse basal line; posterior margin yellow. Fem. Wings above white; anterior, with the tips and marginal oblique stigma black. Posterior, above margined with black; beneath (in both sexes) yellow, with two transverse brown lines. [Papilio Melite. *Fab. Ent. Syst.* 160, 494. *Cramer, tab.* 153. C. D.]

T he remarkable size of the under wings in the male insects of this genus will distinguish them even to a casual observer as forming a natural group. They are all natives of South America, where I discovered nine species. The females differ most strikingly, and have hitherto been mistaken by authors not only for distinct species, but as belonging to different genera.

Their natural situation will be among the *Pieridæ*, with whose general habit they accord.

The female of this species resembles *Pap. Licinia* of Cramer, except in having a short black stigma in the middle of the anterior border of the fore wings, pointing obliquely to the exterior margin. Cramer's insect, however, is the female of another undescribed species in my cabinet.

The under side of the posterior wings in both sexes is the same.

16. *Ismene Œdipodea.*

Generic Character.

Antennæ cylindrical, thickest near the middle, the terminal half subulate; articulations numerous, very short, hardly perceptible. Palpi thick, scaly; frontal side obtuse; lateral sides compressed; the margins externally fringed with hair; the last joint naked, lengthened, nearly horizontal, linear, compressed. Eyes (in the male) very large. Body (in the male) of seven joints, the last with a transverse, slightly emarginate, truncate appendage above; and two obtuse recurved hooks below, concealed by a tuft of hair; the first and last segment shortest. Body in the female with the last joint lengthened and pointed.

Specific Character.

Wings above fuscous, shining blue at their base; posterior margined with orange; beneath clouded with rufous and brown, and a black dot at the base of the posterior wings. Anterior wings (in the male) orange at the base, and a large velvet-like spot of black.

The resplendent and changeable azure blue which ornaments the body and part of the wings in this very singular insect, can be but ill expressed in the figure. It is one of the many new and interesting subjects in entomology discovered in Java by Dr. Horsfield; and by his kindness and liberality I am enabled to add the figures of the caterpillar and chrysalis, which were copied out of a fine series of drawings made in Java under his own eye: they do not appear to differ in their formation from others of this family, although the perfect insect possesses such striking and peculiar generic characters; one of the many facts which prove the impossibility of making the *Larvæ* a primary consideration in forming the genera of Lepidoptera.

This is a rare insect, I have therefore been obliged to leave the generic character imperfect, as the dissection of the mouth, &c. would destroy the specimen. The posterior margin in the wings of the male is sinuated; in the female it is nearly straight; the underside of the wings in both sexes is the same; the anterior pair reddish-brown, paler in the middle; the tip and posterior margin whitish: inferior wings reddish-orange towards the inner margin, with an obsolete central curved band of the same, and a black dot at the base of the inferior wings. The head, palpi, and thorax are margined with orange, less conspicuous in the female.

Our knowledge of the genus *Hesperia* of Latreille (under which the present insect would come) is little more than what was known of *Scarabæus* twenty years ago; nor has Fabricius even noticed one half of the species figured by Cramer. The larva feeds on Gærtnera *Javensis* (*Foliis ovatis, obtusè-acuminatis, caule volubili ramosissimo, ramulis diffusis, deflexis*) a new species, discovered in Java by Dr. Horsfield, who has distinguished it by the above specific character. He informs me the natives give it the name of *Kakas-rambat*, which last word signifies twining or trailing. In the inflorescence and fruit it differs not from *G. racemosa*.

17. *Bulimus zonatus.*
Zoned Bulimus.

Generic Character.

See Plate 4.

Specific Character.

Shell smooth, conic, of five volutions, the last somewhat distorted; white, with two unequal fer-rugineous bands; body whirl rufous, with two white bands. Aperture white.

A small though very elegant shell, seldom seen in Collections; nor do I find such a description of it as will identify the species. One figured by Martini, at *tab.* 134, *fig.* 1215, comes near it, but differs sufficiently for a specific distinction.

Its precise locality is unknown: a fine specimen exists in my father's collection, who thinks it came from the East Indies; and this is the only one I have yet seen.

The aperture is more round than ovate, and is less than one half the total length of the shell; the outer lip much reflected, and the transverse bands on the spiral whirls nearly obsolete.

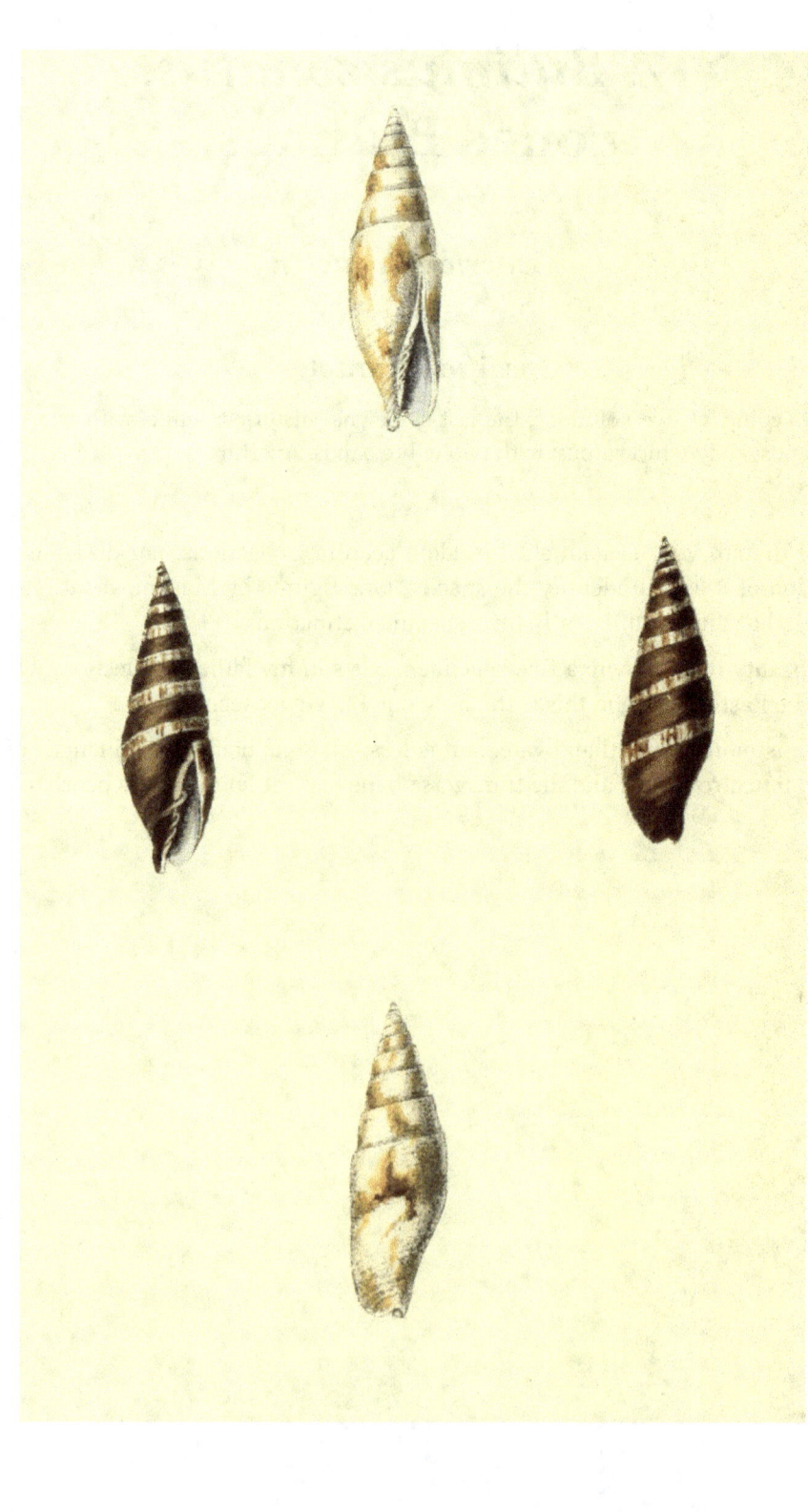

18. *Mitra.*
Mitre

Generic Character.

See Plate 23.

Specific Character.

Shell nearly smooth; upper margin of the volutions prominent; basal whorl contracted in the middle.

Contracted Mitre —upper figure.

An undescribed species, for the loan of which I am indebted to Mr. G. Humphrey, of Leicester-street, whose knowledge as a collector, and integrity as a dealer, have gained him respect and confidence through a long life: and it is no less singular than true, that many genera of modern authors, now universally adopted, were formed by him near twenty-five years ago (under different names) in the *Museum Calonianum*, printed in 1797.

Shell one inch and a half long, and smooth; the base and spire with faint remote grooves; the spiral whorls are scarcely convex, and their upper margins prominent. Outer lip thick, effuse, slightly reflected below, and contracted above. Pillar five-plaited; colour yellowish white, with two or three waved longitudinal bands of orange, and a few others broad and remote on the spire. A finer specimen I have since seen with Mr. Humphrey's was one inch three quarters long, the ground-colour pure white, the aperture orange, and the bands rich orange-chesnut.

New Holland Mitre—lower figure.

Shell very smooth. Spire elongated, chesnut-brown; base of the spiral volutions with a whitish band, which is central on the basal whorl. Pillar four-plaited.

Dead shells of this new Mitre were received from Van Dieman's Land by Mr. Humphrey: it is perfectly destitute of striæ, excepting a few faint ones at the base: the mouth, which is smooth inside, appeared in the few specimens he had, to be unformed; it is, however, sufficiently distinct from any other.

19. *Tinamus Tataupa.*
Tataupa Tinamou.

Generic Character.

Bill moderate, depressed, broader than high, tip obtuse, back broad. Nostrils lateral, medial, ovate, expanded and open. Feet four-toed, cleft; hind toe very short. Tail none or very short, concealed by the rump-feathers. Wings short. Generic Type *T. rufescens.* Latham.

Specific Character.

Tinamou with the body above dusky-rufous, immaculate. Head and neck dusky-black; chin white; throat, neck and breast, cinereous; body beneath whitish; vent and flanks rufous or black, the feathers margined with white. [Tinamus Tataupa. *Temminck Pig. et Gall.* iii. *p.* 590 *et* 752. *Gen. Zool. vol.* xi. *part* 2. *p.* 416.]

T he Tinamous are entirely confined to the new world, where they seem to hold the same scale in creation which the Partridges do in the old continent. Our knowledge of these singular birds has been much increased by the writings of Professor Temminck, who has described twelve species. The present bird is nearly the smallest of its family: I found it only once in the interior of Bahia in Brazil, where it must be very rare, or frequent to particular districts only. Though differing in some respects from the description of Temminck, I am inclined to consider it merely as a variety.

Total length (excepting the legs) eight inches and a quarter. The bill is one inch one line long from the gape, and, with the irides, is red. The head and neck above blackish cinereous; the crown much darker and tinged with brown, the rest of the upper plumage uniform reddish-brown; the edges of the wing-covers tinged with pale cinereous; the spurious wings and quills greyish-brown; the chin is white, changing on the throat, neck, breast and their sides to a pale lead-colour, which, on the body, again becomes white; the feathers on the flanks are blackish or rufous, beautifully margined all round by white, with another internal mark of the same kind; those on the vent are similarly marked, but on a pale rufous ground; the thighs are rufous-white; the under tail-covers rufous, marked by narrow undulated concentric lines of black, the ends whitish. The length of the legs (from the knee to the base of the middle toe) one inch two lines, and from that to the tip of the claw one inch. Legs blueish-purple. Hind toe very short, and elevated above the ground.

20. *Picus Braziliensis.*
Brazilian Woodpecker.

Generic Character.

See Plate 14.

Specific Character.

Olive Woodpecker: beneath fulvous, with transverse blackish bands. Head sub-crested, above red, the sides with olive, yellow, and red streaks. [P. Braziliensis, Swains. in Wern. Trans. 3. p. 291.]

A new species of this already extensive family, inhabiting the interior of Brazil in the province of Bahia, where I met with it but once. It was, I believe, first described in a paper I sent to the Wernerian Society some time ago: the figure is less than the natural size.

Total length nine inches. Bill not quite an inch, and blackish. Irides yellow. Head slightly crested; the whole upper part crimson. Orbits and cheeks olive-brown; beneath this a narrow line of tawny-yellow begins at the nostrils and passes down the sides of the neck; next this is a similar stripe crimson on the jaws and olive beyond, leaving the chin and throat in front yellowish; the plumage above is tawny-olive. Quills black, within edged with rufous: all the under parts tawny-yellow, transversely banded with blackish lines; inner wing-covers yellowish. Tail three inches and a half long, the feathers black, unspotted, and tinged at their base with olive. Feet and claws lead-colour. The neck is very slender. The only one I have yet seen was a male.

21. *Procnias hirundacea.* Swallow Fruit-eater.

Generic Character.

Bill short, triangular, base very broad, dilated, towards the end contracted; both mandibles notched, the margins bent inward; upper mandible slightly curved and carinated above; lower mandible straight and shortest. Nostrils broad, basal, nearly naked, the aperture much nearer the tip than the gape of the bill. Tongue very short, narrow. Mouth very large, opening beneath the eye. Feet formed for perching. Wings moderate. Generic Type *Carunculated Chatterer.* Latham.

Obs. MM. Temminck and Lagier had just before us, and without our knowledge, published this bird under the name of Procnias Ventralis, (Pl. 5.) by which name in right of priority it should stand in the system.

Specific Character.

(Male) blue; front, throat, and temples black; middle of the body beneath white, the sides with blackish transverse striæ.

(Female) green; chin and temples grey; body beneath yellowish, transversely striated with dusky-green.

The birds of this genus are remarkable for the enormous width of their mouths, which in some species exceeds that of the Swallow family, thus enabling them with ease to swallow the large berries of the *Melastomæ* and other tropical shrubs, on which they alone subsist; not on insects, as Cuvier asserts. Although in the construction of their bills they perfectly resemble the Swallows, their wings are not formed for long or rapid flight; and their feet are much stronger, and calculated for searching among branches for their food, in which situations I have frequently seen them. The term "*pedes ambulatorii,*" or walking-feet, is applied too generally, and should be confined to the gallinaceous and Pigeon tribes.

This genus was formed by Count Hoffmansegg, and the present is the smallest species known: our figure is of the male bird. Total length about five inches and a half. The bill from the angle to the tip measures seven lines; but from the nostrils only three lines and a half. The middle of the body, vent, and under tail-covers in the male are pure white; in the female yellowish, with a line of olive-green down the middle of each shaft; the quills, wing-covers, and tail-feathers are black, margined in the male with blue, and in the female with green: the tail is slightly forked. The nostrils round and bare; the base of the bill has a few weak setaceous hairs. The legs resemble the true Chatterers, having the outer toe rather longer than the inner, and attached to the base of the middle.

This is a scarce bird, apparently not hitherto described; I met with it only three times in Bahia; but it appears more frequent in the southern provinces of Brazil, specimens having been sent me from Minas Geralis and Rio de Janeiro.

22. *Terias Elvina.*

Generic Character.

Antennæ short, the club somewhat truncate and compressed. Palpi very short, curved, hardly projecting beyond the head, closely compressed on the tongue, entirely covered with close imbricate scales, the tip naked. Body elongated, slender, in the male six-jointed, the last with two approximating incurved hooks; valves broad, thickened, truncate, and hooked. Wings in both sexes alike, broad, obtuse, rounded, very entire. Generic Type *Papilio Hecabe.* Linnæus.

Specific Character.

Wings sub-diaphanous, pale sulphur; beneath immaculate. Anterior, above with a black marginal tip; posterior (in the male) with the fore-margin gibbous at the base. Female ——? [Pieris Elvina. *Godart in Encycl. Method, p.* 158. *no.* 67.]

T his is one of the smallest of Butterflies, and from the extreme delicacy of its form seems to sanction with truth the poetic idea of living "but for a day." It is found in Brazil, inhabiting only the deepest forests, as if fearful its little life would be endangered by the scorching rays of a tropical sun: in these sombre shades it is seen to fly slowly and feebly near those spots where a ray of the sun has partially entered the thick canopy of foliage above, which is frequently fifty or sixty feet from the ground.

The genus I have now placed it in belongs to the *Coliadæ*, and appears to connect that family with the *Pieridæ*: their distinctions are obviously marked and very constant in all the species I have yet seen, and which are tropical: of these, seven I discovered in Brazil; three or four more are natives of the southern extremity of North America; and Dr. Horsfield has four or five from Java. I know of none from Africa. Their size in general is very small.

I think this species is the *Pieris Elvina* of Godart; although the insect he mentions as the female is in reality that of his *Pieris Neda.* The true female I have never seen; I suspect it will want the gibbous curve on the hinder wings of the male, which sex is, indeed, not common, and is generally much smaller, and sometimes half the size only, of the figure.

Papilio Nicippe of Cramer (tab. 210. fig. C. D.) strictly belongs to this genus, though placed in that of *Colias* by Godart, as well as his *Pieris Agave, Hecabe,* and doubtless many others not now before me.

23. *Mitra vittata.* Ribbon Mitre.

Generic Character.

Shell unequally fusiform; spire lengthened, attenuated; outer lip simple not toothed within. Columella plaited.

Divisions.

I.   Aperture narrow, linear, above angulated, below a little contracted.

Mitræ vulpecula, plicata, &c.

Obs. Shell generally longitudinally plaited, equally fusiform; outer lip smooth, slightly waved; top of the inner lip much thickened within; throat striated.

II. Aperture above pointed, below narrowed, externally curved.

Voluta mitra-abbatis. Chemnitz, &c.

Obs. Shell generally with an elongated spire, the aperture below narrowed; upper syphon or channel small or wanting.

III. Aperture above pointed, externally straight, below rounded, widened or effuse. *Mitræ papalis, episcopalis,* &c.

Obs. Shell generally smooth, the base thick and truncated; margin of the outer lip crenated; throat smooth. The smaller shells of this division connect the genera *Mitra* and *Colombella* (Lamarck).

Specific Character.

Shell narrow, base cancellated. Spire with carinated plaits, the interstices with slender, crowded, transverse grooves. Pillar of four plaits; throat with four to five remote striæ.

This superb shell is figured from a matchless specimen brought home by that illustrious and lamented patron of science, the late Sir J. Banks, from the Pacific Ocean: it is now, together with his entire collection of shells and insects, in the Museum of the Linnæan Society.

It is of great rarity, and the present specimen far exceeds in size any I have yet seen. A very perfect one exists in my father's collection which measures only two inches one line long: it differs slightly in wanting the lower white band and its inferior border: there is also an additional small plait between the second and third, a variation not uncommon in the Linnæan Volutes, and which lessens the importance of this character as a specific distinction.

It is unfigured, and I believe undescribed, unless perhaps in Solander's MSS. In its small state it may have been overlooked as one of the numerous varieties of *M. vulpecula*; but the sharp angulated plaitings, the cancellated base, and the numerous faintly-grooved lines on the spire, as well as the more slender and lengthened form, will at once distinguish it: its colours also are very striking and dissimilar.

24. Conœlix.

Generic Character.

Shell coniform. Spire very short. Outer lip simple. Columella or pillar plaited. Aperture linear, narrow, longer than the spire. Generic Type *Conœlix lineatus.*

Marbled Conœlix—upper figures.

Shell with remote capillary transverse striæ. Spire slightly produced, acuminated; the whorls with a central indented line. Outer lip crenated.

The rare little shells composing the group I have now formed into the genus *Conœlix,* seem to have escaped the observation of modern systematic writers. They form a beautifully defined link connecting the Cones with the Volutes, strictly so termed, and their generic characters seem to be very constant and clear. The present species varies more or less in the regularity of its tessellated markings. The inside of the mouth is brown, and the pillar has five plaits. Several specimens are in the Banksian Cabinet, from the Pelew Islands. The figures are cnlargcd to one half more than the natural size.

Lineated Conœlix—middle figures.

Shell smooth, whitish, with transverse capillary fulvous lines. Spire depressed, the apex prominent. Pillar six-plaited.

Figured of the natural size. The volutions of the spire are somewhat convex; the coloured lines are not indented. Inhabits the South Seas?

Punctured Conœlix—lower figures.

Shell cream-colour, with capillary transverse striæ, which are minutely punctured. Spire short. Pillar five-plaited.

Inhabits Otaheite: from the Banksian Collection. The figures are on the same scale as *C. Marmoratus.*

These are the only three species which I have myself seen. Another is figured in *Chemnitz* x. *tab..*150. *fig.* 1415 and 6. Mr. Humfreys informs me he has seen at different times five or six others, all of a small size.

25. *Procnias Melanocephalus.*
Black-headed Berry-eater.

Generic Character

See Plate 21

Specific Character.

Olive-green, beneath yellowish, with dusky transverse striæ. Head entirely black.

Another new and very rare bird of this singular genus, inhabiting, like all the other species, the tropical regions of America. I met with it in Brazil but twice in the forests of Pitanga, not far distant from Bahia; and my hunters were at a loss for its name, never having seen it before: the eyes in the fresh bird are of a beautiful crimson.

Its total length is nine inches and a quarter; the bill is nine lines from the gape to the tip, and four from the base of the nostrils, at which part the bill is not so proportionably broad as in the Swallow Berryeater (plate 21): the colour blueish-black, paler at the base: the whole head, sides, chin, and part of the throat are black, the feathers of the crown a little lengthened and pointed, giving a slight appearance of a crest: the wings and tail are dusky-black on the inner shafts and green on the outer; the whole of the upper plumage olive-green, and of the under pale greenish-yellow crossed with short dusky transverse lines from the breast downwards; under wing and tail-covers the same. Tail four inches from the base, slightly divaricated, and of twelve feathers. Wings four inches and a half, the first quill very short, the third, fourth and fifth of equal length. Legs black.

This was a male bird: the female I have not seen.

26

26. *Alcedo azurea.*
Azure Kingsfisher.

Generic Character.

Bill very long, straight and attenuated, higher than broad, compressed the whole length, both mandibles carinated, the margins slightly bent inwards. Nostrils basal, covered by a membrane; the aperture linear, oblique, and naked. Tail mostly very short. Feet gressorial, inner fore-toe small or wanting. Generic Type Common Kingsfisher. Lath. Bewick, &c.

Specific Character.

Body above, sides of the head and neck shining mazarine blue; beneath rufous; chin and throat whitish; wings blackish; inner fore-toe wanting. [Alcedo azurea. Azure Kingsfisher. *Lath. Synop. Suppl.* ii. *p.* 372.] [Alcedo azurea. *Lewin's Birds of New Holland, fasc.* i. *pl.* 1.] [Alcedo Tribrachys. Tridigitated Kingsfisher. *Shaw in Gen. Zool.* viii. 1. 105.]

The Kingsfishers have such a general similarity of form, that the most casual observer is able to distinguish them: a very long straight bill, short wings, and (in general) a shorter tail with very small legs, are the prominent distinctions of such as are usually seen; and the richness of plumage that generally pervades them cannot be better exemplified than in our own beautiful species, the common Kingsfisher, not unfrequent in many parts of England.

These birds, hitherto placed in systems under one genus, nevertheless contain two distinct groups differing materially in the construction of that primary organ of supporting life, the bill; and in their physical distribution, or the countries they respectively inhabit, two most important considerations in the natural arrangement of animals under the present elevated views of the philosophic zoologist, with whom the study of Nature consists no longer in the study of words, the retention of names, or even the accurate description of species.

These considerations have induced me to form these birds into two genera, the definitions of which are now given: those retained under the old genus of Alcedo appear to be scattered (though sparingly) in every part of the old and the new world. Their bills seem formed for swallowing their food more in an entire state, similar to the Herons. In each of these genera one species exists with only three toes, a remarkable circumstance, which in an artificial system would endanger their being united in a separate genus; but which, from the remarkable smallness of the inner toe in all the other species, cannot I apprehend point out any peculiarity either in their habit or economy: and this opinion I find is likewise entertained by Professor Temminck.

Total length seven inches and a quarter. Bill from the gape two inches one line, the upper mandible rather longest, and both with a slight appearance of a notch; the colour black. All the upper plumage, as well as the sides of the head, ears, and stripe beyond, fine ultramarine blue, more vivid on the rump and tail-covers, and duller on the tail, wing-covers, and lesser quill-margins; front blackish; from the nostrils to the eye a whitish line, and from the ears on each side the neck a whitish stripe, which almost forms a collar round the nape. Quill-feathers sooty black. All the under parts orange ferrugineous; throat and belly nearly white. Tail very short, nearly hid by the upper covers. Feet red, claws black. The inner fore-toe wanting, but a slight rudiment of it exists in my specimen.

Since writing the above, I find this bird is figured and described in a beautiful work commenced by Lewin on the birds of New Holland, which Mr. Brown, the learned possessor of the Banksian library, pointed out to me. I believe but a few copies are known. Lewin observes, "it inhabits heads of rivers, visiting dead trees, from the branches of which it darts on its prey in the water beneath, and is sometimes completely immersed by the velocity of its descent."

Dr. Latham has very well described it, but quite overlooked the construction of the feet.

27.

27. Halcyon collaris.
Collared Crabeater.

Generic Character.

Bill very long, straight, thick, the base broader than high; the sides tetragonal; upper mandible very straight, the base rounded; under mandible beneath carinated and recurved, the margins covered by those of the upper. Nostrils basal, covered by a membrane, the aperture naked, linear and oblique. Tail mostly moderate. Feet gressorial: interior fore-toe small or wanting. Generic Type Crabeating Kingsfisher. Latham.

Specific Character.

Greenish-blue. Body beneath and nuchal collar white. *H. viridi-cærulea; corpore subtus, lunulaque cerviculi albis.* [Greenish-blue. Body beneath and nuchal collar white.] [Alcedo collaris. *Latham Index Ornith.* i. 250.] [Sacred Kingsfisher, *Var.* D. *Latham Syn.* ii. *p.* 623.] [Collared Kingsfisher. *Gen. Zool.* viii. i. *p.* 80.]

Referring to the observations we have already made on Kingsfishers generally, it will be only necessary to observe, that the species now formed into the genus Halcyon appear entirely excluded from the American continent: their bills are much stronger, thicker, and more rounded than the genuine Kingsfishers, and the under mandible beneath invariably carinated and curving upwards. One of them (the Alcedo Senegalensis of Latham) is known to feed on crabs, the breaking and disjointing of which this structure seems admirably calculated to accomplish; and although some authors mention insects also as their food, I apprehend it is only in the absence of other larger prey more suited to the construction of their bills.

Total length eight inches and a half. Bill two inches three lines from the gape, and one inch three quarters from the nostrils; upper mandible and margin and lip of the lower, black, the rest yellowish-white. The general plumage above is pale and changeable greenish-blue, the green predominating on the scapulars, head and tail; the upper part of the neck is crossed by a white collar, separated from the green of the head by a narrow margin of black, which passes on the ear-feathers round the nape; a narrow whitish line runs from the nostrils to the eyebrows, and another very short one is beneath the eye; the whole of the under plumage white. Quills black edged with blue, the second, third and fourth equal and longest. Wings four inches and a quarter. Tail even, near three inches long, above blue-green, beneath black. Feet dusky; middle and outer claws much longer than the leg.

Inhabits Java and other parts of India, and is I believe unfigured. The line at the bottom of the plate is on the scale of an inch.

Since writing the above, Temminck's new edition of the Manuel d'Ornithologie has just reached me, in which I perceive he has continued the birds of this genus under that of Alcedo, observing that their plumage is always shining, and that he can find no characters for their geographic distribution: yet, notwithstanding the opinion of this eminent ornithologist, a close attention will I believe prove, first, that no species of Linnæan Alcedo bearing the characters of Halcyon have yet been discovered as natives of America; and secondly, that species of genuine Alcedo will be found with plumage quite devoid of any bright or shining colours. One or two exist in my own cabinet, but to which I cannot now refer.

The situation of Halcyon will be between Alcedo and Dacelo; from the last of which it is distinguished by its perfectly straight, acute, and entire upper mandible, which, on the contrary, in Dacelo is notched, the tip bent and obtuse.

28. *Hesperia Haworthiana.*
Haworth's Hesperia.

Generic Character.

Antennæ moderate or elongated, straight, slender, the club nearly terminal, short, thick, cylindric, ending in an abrupt, short and pointed hook. Palpi compressed, incurved in front of the head, the sides convex or angular; the last joint erect, pointing vertically. Wings when at rest erect.

Divisions.

I.   Palpi broad, very compressed in front. Antennæ short, the club very thick.

II. Palpi nearly square, very thick. Antennæ elongated.

III. Palpi with the last joint lengthened, slender. Antennæ moderate.

Generic Type *Hesperia Comma* of Authors.

Specific Character.

Hesperia (Div. 2.). Wings above blackish-brown, the base shining blue; anterior with a medial hyaline band; posterior beneath brown, with two longitudinal yellow-green lines. Legs brownish-orange.

The celebrated Latreille, the father of modern Entomology, has well observed, that the immense number of insects crowded together in the genus *Hesperia* contain many natural genera, but which the paucity of species generally found in cabinets prevents us from discriminating. Having for a long time paid attention to this family, and possessing near 300 species in my own cabinet, I have had the opportunity of attempting their elucidation; and the above generic character is applied to those insects only which I propose considering genuine species of the genus *Hesperia*, and which will comprise near 170 species.

I have named this new, undescribed and very rare insect, in honour of my esteemed friend A. H. Haworth, Esq. F.L.S., &c., well known by the benefits his writings have conferred on the sister sciences of entomology and botany. The only two insects I ever saw of this species I captured in the southern part of Brazil.

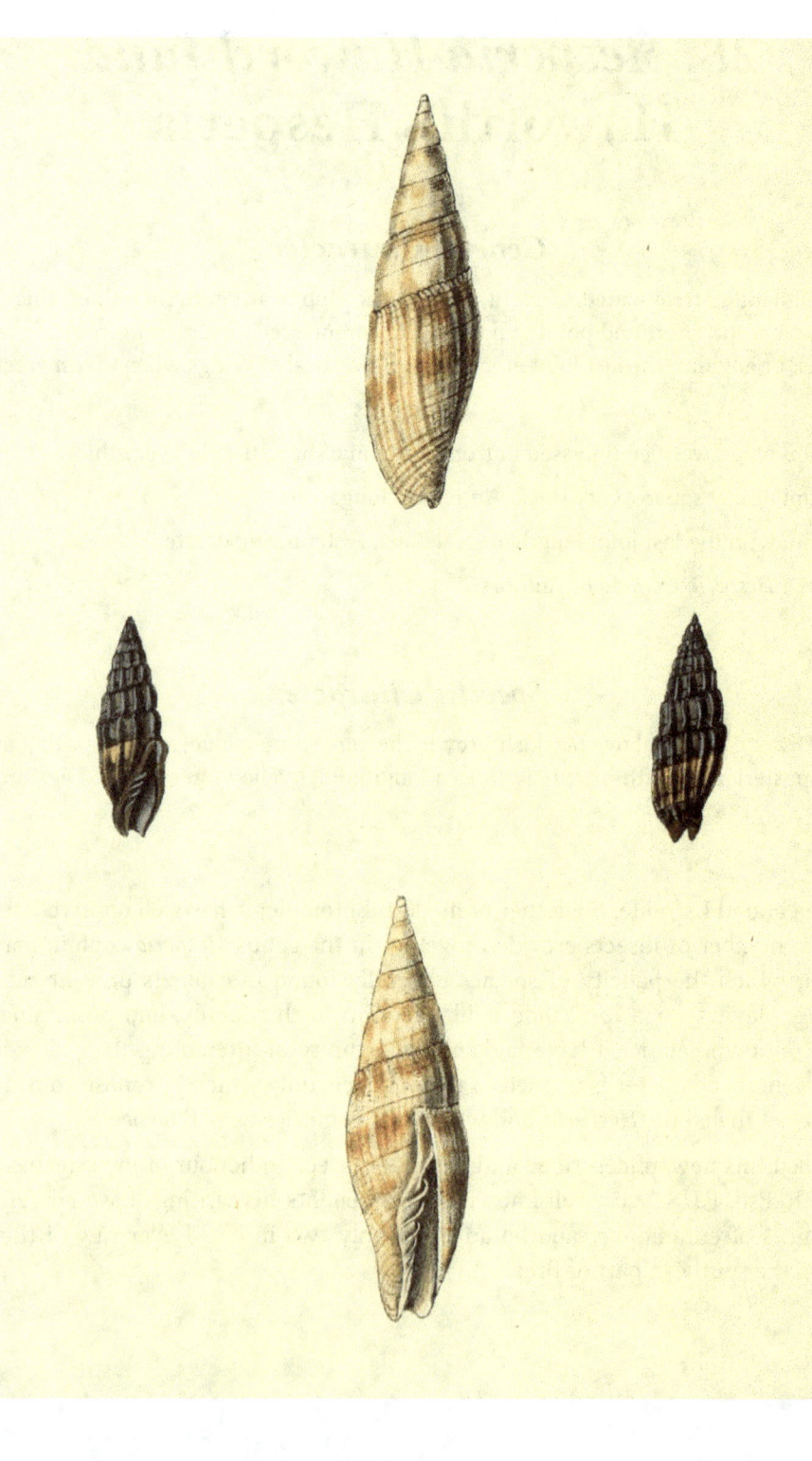

29. *Mitra.*
Mitre

Generic Character.

See Plate 23.

Specific Character.

Basket Mitre—upper figure.

Shell fusiform, cancellated, the longitudinal striæ thickened; spire and aperture of equal length; pillar five-plaited; spire nearly smooth.

Another undescribed species of this elegant family, and of great rarity, in the private collection of Mr. G. Humfreys. The whole of the body whorl and commencement of the spire is cancellated. The longitudinal striæ are crowded, thickened, and slightly elevated, giving a crenated appearance to the suture: the transverse striæ slender, and filling up the interstices. The spire is nearly smooth and a little bent: the ground colour very light orange, with three darker interrupted bands on the body: whorl separated by two slender lines of the same colour; the spiral whorls have only two bands and a line between; the upper margins slightly compressed on the suture; the outer lip within smooth.

Ribbed Mitre—middle figures.

Shell with elevated, longitudinal, obtuse, entire ribs, the interstices smooth, the base granulated; spire lengthened; pillar four-plaited; aperture short.

Equally rare, and from the same collection as the preceding. In habit it approaches nearest to *M. exasperata* of Chemnitz, but has not the ribs angulated or their interstices striated, and is much more narrowed at the base than in that shell, which I have seen: the outer lip is also smooth; the inside strongly striated. This shell was formerly in the collection of Mr. Keate, the elegant author of the "*Sketches from Nature.*"

30. *Achatina marginata.* Marginated Achatina.

Generic Character.

Shell ovate, or oblong-ovate; spire elevated; mouth nearly oval. Columella smooth, simple, truncated. Outer lip thin; inner lip entirely inflexed. Umbilicus none. Generic Type *Bulla Achatina.* Linn.

Specific Character.

Shell ovate-oblong, with irregular ferrugineous stripes; spire obtuse at the top, of five volutions; the suture depressed, with a marginal indented line. [*Lister* 579. *fig.* 34. *Gualt. pl.* 45. B. *Knorr, vol.* iv. *tab.* 24. 1. (badly coloured.)]

The largest shells hitherto discovered as inhabiting the dry land belong to this genus, instituted by the celebrated Lamarck, but still divided by the strict followers of Linnæus between the *Bullæ* and *Helices*, with a singular infelicity of even artificial arrangement. The simple characters peculiar in a greater or less degree to all, will readily distinguish them; and I apprehend most of the species of the first division (which includes the present) will be found to inhabit only the African continent, while *Bulla virginea* and the smaller shells placed in the second division are found principally in the new world; where also two or three gigantic species of *Bulimus* occupy the place of the larger African *Achatinæ.*

Of these, the shell now figured is one of the rarest, and has hitherto been overlooked as a variety of the Linnæan *Bulla Achatina*; the colour of both is subject to much variation; but this will be found at best a most indecisive and vague character for specific distinction when unaccompanied by others more important and connected with the formation of shells. I have therefore not hesitated in making this a distinct species, from having had the means of examining at different times near twenty specimens, all of which presented the following characters. Spire of five whorls, the last or terminal one very small and flattened; the apex obtuse; the suture depressed, as if flattened on the shell, and margined by one or sometimes two indented lines, parallel, and at the top of each whorl. In the colour of its mouth it varies in sometimes having a tinge of rose-colour at the base and top of the spire, but the mouth is more generally white. The body whorl is more or less ventricose; the outer lip is a little reflected, and the whole shell, when full grown, much thicker and heavier than any of the other species. The epidermis is yellowish-brown, beneath which the shell is nearly white, beautifully marked with broad remote stripes of chesnut, with others more slender (and sometimes broken into spots) between. I have another specimen which agrees tolerably with Lister's figure in being more than usually ventricose, and which I think is accidental. The only constant variety appears to be that figured by Knorr, ii. tab. 3. fig. 1. having the spire entirely rose-colour.

The marginal line and the correct number of whorls in the spire are well expressed in the figures of Lister, Gualtieri and Knorr. The first of these figures is accidentally more ventricose; the second, like all the other figures of Gualtieri, is defective at the apex; and Knorr's I suspect has been outrageously coloured from the real pink-mouthed *Achatina*.

It inhabits the coast of Guinea; and I am informed the animal is eaten by the natives.

31. *Phibalura cristata.*
Crested Shortbill.

Generic Character.

Bill very short, triangular, broader than high; upper mandible above slightly curved and carinated; lower mandible straight, both notched. Nostrils simple, basal, roundish, entirely concealed by thick-set incumbent feathers. Mouth large, opening beneath the eye. Wings pointed; spurious quills none. Tail elongated, forked, of twelve feathers. Feet formed for sitting; the fore-toes equally cleft and slightly connected at their base.

Specific Character.

Above black varied with yellow; beneath white, with transverse black bands; chin yellow. Head crested, the feathers rufous, varied with black. Wings, and elongated forked tail raven-black, immaculate.

For this beautiful and extraordinary bird I am indebted to Miss E. Yeates, of the Dingle near Liverpool, who received it from South America. Its general habit clearly points it out as belonging to the *Baccavoræ* or Berryeaters, apparently connecting the genera *Procnias* and *Pipra*, where Temminck with much judgement has also placed it, in the new edition of his *Manuel d'Ornithologie* just received, and before reading which I had considered the genus as unpublished.

The total length is nine inches, of which the tail occupies four and a half. The bill is whitish, and is remarkably short, measuring only three lines from the nostrils to the tip, but three quarters of an inch from the angle of the mouth, which opens just under the eye: the plumage is singularly variegated: the crown of the head is furnished with a crest, which, when not elevated, is scarcely seen, and appears a deep glossy black mixed with grey and rufous; but when erected it is very conspicuous, and all the feathers are bright rufous tipt more or less with black; the upper sides of the head grey, the lower part and ears deep-black; the neck above is greyish-white, with blackish transverse lines: the back, scapulars, rump and tail-covers are varied transversely with olive, shining black, and bright yellow, each feather being olive at the base, black in the middle, and yellow at the tip. Beneath the feathers of the chin and part of the throat are somewhat lengthened, semi-setaceous, and of a bright yellow; the neck and breast are white, with two transverse lines of deep black on each feather; these lines diminish, and are broken into spots on the body, and nearly disappear on the vent: the edges of the breast-feathers are tipt with yellow, which colour increases downwards on the vent and tail-covers, which latter are entirely yellow. The wings are four inches long, uniform deep black with a blue gloss, much pointed, and calculated for rapid flight. Tail the same colour, the exterior basal margins olive: all the feathers are narrow, pointed, and gradually lengthening, the middle pair being two inches three quarters longer than the outer pair, which exceed those next them by an inch. The feet are very pale yellow, and three-quarters of an inch from the knee to the claws, the three foremost of which are equally connected together (though slightly) nearly as far as the first joint; the outer and inner toes equal, and rather shorter than the hind-toe: claws slender and much compressed.

Whether this species is the same as the one mentioned by Temminck as existing in the French Museum under the name of *P. flavirostris*, it is quite impossible to say, as the description of that bird has never been published. This leads me to notice a custom several naturalists of the present day have lately adopted, of publishing names, and names only, of new or undescribed animals, which they then wish to be considered as permanently fixed, and as having thus secured to themselves all the merit of first describing. Now this at best is but a surreptitious path to fame, and in many instances bears the appearance of originating in a petty vanity, quite beneath the dignity of true science: it is easily fixing a name to an object which we have not before seen, or suspect may be new, without the trouble of investigating authors and comparing synonyms: the name may remain, but if it should afterwards be discovered as hasty and erroneous, its author is in no way amenable to the opinions and criticisms of others, for they cannot discover such mistakes when no clue is given them beyond a name, which may frequently be applicable to half a dozen species. If, on the other hand, the object is really new, the scientific world is still in the dark, for without a description the name conveys nothing. Besides this, it has a tendency to deprive those writers of their well-earned merit, who undergo the laborious but necessary investigation of books, the examining and comparing of specimens, and the construction of sound characters previous to their publishing a new addition to the great volume of Nature. Against this *scientific monopoly* a stand should be made, and all names either of families, genera, or species should be totally rejected, unless their meaning is clearly defined. Let those who run the race, receive the wreath; and not let it be snatched from the winning-post by another, who jumps from behind and claims it as his own.

On a careful examination of my specimen, I find the nostrils are not covered by a membrane, as observed by Temminck, but are open, obliquely and ovately round, and a narrow rim round the margin. That excellent ornithologist likewise remarks that the first and second quill-feathers are the longest; but my bird (which, however, is in full plumage) has the first and third of equal length and shorter than the second, which is longest. These nice distinctions lead me to suppose the species from which his generic character was taken, is distinct from this.

32. *Psaris Cuvierii.* Cuvier's Psaris.

Generic Character.

Bill strong, thick, conic, the base rounded, towards the top slightly compressed, the top convex, not carinated; both mandibles notched, the tip of the upper hooked. Nostrils basal, simple, round, situated near the margin, the base with a few short incumbent setaceous feathers. Feet simple, the three fore-toes equally cleft. Spurious quills none. Tail short, of twelve equal feathers. Generic Type *Lanius cayanus.* Linn., Lath., &c.

Specific Character.

Olive, beneath whitish; crown black; nape, sides of the head and neck pale cinereous; breast, sides, and under wing-covers yellow.

The genus *Psaris* was first instituted with great propriety by Cuvier; and before the discovery of the species now made known, was supposed to consist of only one, the Cayenne Shrike of Latham, which with the present bird (named in honour of the first zoologist of the age) is found in Brazil. The figure is nearly of the natural size.

Total length five inches and a half. Bill blueish, three quarters of an inch from the angle of the mouth, and four-tenths from the nostrils, which are ovately round, rather large, and simple, being entirely devoid of an external membrane, but the base is partially covered with small thick-set, short, setaceous feathers; between the eye and base of the bill are a few weak and short hairs; the upper part of the head, as far as the nape, is capped by deep-black, having a blueish gloss: between the nostrils and the eye, as well as on the chin and throat, the colour is white, which changes to a pale cinereous grey on the sides of the head and round the neck; the ears at their base and margin of the eye tinged with yellow; the rest of the upper plumage yellowish-olive. The under plumage on the lower part of the neck and breast, the sides, and the inner wing-covers are clear yellow, and from that to the vent white. Wings two inches long, the quills brown, margined externally with olive and internally with yellow; the first and second quill progressively shorter than the third and fourth, which are of equal length. Tail short, slightly divaricated; olive, with whitish marginal tips. Legs blueish-black; the three fore-toes are equally cleft, but a membrane will be found connecting them equally at the base nearly as far as the first joint.

Temminck must be mistaken in giving as a generic character to this genus, that the external toe is connected to the middle one as far as the first joint, and the inner toe cleft to the base; at least such is not the case either in my specimens of this bird or in those of the Cayenne Shrike: and they have been carefully relaxed in warm water, the best method of ascertaining such peculiarities.

33. *Tamyris Zeleucus.*

Generic Character.

Antennæ arcuated; the club terminal, thick, linear, obtuse; more slender and attenuated in the female. Palpi compressed convexly on the front of the head, meeting above the tongue; the last joint very minute, thick, obtuse, approximating and bent forward. Wings short, when at rest horizontally divaricated.

Specific Character, etc.

Wings uniform blueish-black, with a slender white margin. Head and top of the body bright red. [Hesp. Zeleucus. *Fab. Ent. Syst.* 3. *pt.* 1. *p.* 346. *no.* 317.]

Obs. *Donovan's Indian Insects*, where that author has figured it by mistake as a native of India.

This insect is the most common (although hitherto unfigured) of a striking natural group belonging to the *Hesperidæ*; it has therefore been selected as the best example for the genus I have now formed them into. I have not seen more than twelve or fourteen species, and these were all from different parts of South America, to which I have no doubt the genus is exclusively confined. The club of their antennæ is very thick, obtuse, and without any terminal hook. The bright red at the end of the abdomen (improperly called by Fabricius the tail) is most conspicuous in the female, which is also larger and having the wings more obtuse, of which the upper and under surfaces are both alike.

The insects of this family fly with amazing rapidity (as is shown by the thickness of their thorax, and the sharpness in the make of their wings), generally frequenting openings of thick woods and alighting on leaves where the sun strikes: I seldom saw them on flowers. Their wings when at rest are half expanded in a horizontal direction. Their metamorphosis is unknown.

This individual species is scarce in the northern parts of Brazil, but common in the southern provinces.

34. *Colias Godartiana.*
Godart's Colias.

Generic Character

See Plate 5

Specific Character.

(Female) Wings fulvous-yellow; anterior above with the outer margin and round central spot black, which beneath is silvery rufous and three-cleft; posterior beneath each with two silvery spots margined with black, one of which is quadrangular. Palpi lengthened.

An inspection of a vast number of insects of this genus, with the possession of nearly all the species noticed by authors, convinces me that the insect now figured is perfectly distinct from any other. It is in the cabinet of Mr. Haworth, who obligingly lent it me for comparison and description, and is the only individual I have hitherto met with. The prolongation of the palpi, which is even more obvious than in *C. Statira*, is alone a specific distinction; and the form of the spots both on the upper and under side differs very much in character from that insect, with which it has the most affinity. It may be the *Papilio Drya* of Fabr. (omitting his references); but his description, whether intended for this insect or any other, is so vague that I can see no advantage in retaining it. Of the two bright silver spots beneath, one is oval, the other larger and quadrangular.

I have named it in honour of M. Godart, the intelligent coadjutor of M. Latreille in the entomological part of the *Encyclopédie Méthodique*.

35. *Mitra bifasciata.* Double-banded Mitre.

Generic Character.

Specific Character.

Shell smooth, uniform chesnut-brown, with two narrow yellowish bands on the basal whorl, and one on the spire; aperture smooth. [Voluta caffra. *Martini* iv. *tab.* 148. *fig.* 1369.] [Voluta caffra. *Knorr. vol.* v. *tab.* 19. *fig.* 4, 5.] [Voluta caffra. *Seba Pl.* 49. *fig.* 21, 22, 41.]

This most elegant shell has been figured from one of the specimens that belonged to the late Mr. Jennings, who was well known to spare neither expense nor assiduity in procuring the most select and matchless specimens of every species; so much so, indeed, that such as are known to have been in his possession generally bear a higher price. One of these is now in my father's cabinet, the other in that of Mrs. Bolton, of Storr's-hall, Windermere. I have seen both, and they appear equally fine.

I cannot help considering this as a distinct species from *Mitra caffra* (*Voluta caffra* Linn.), with which it has hitherto been placed only as a variety: it is much larger, the volutions more convex, but compressed on the suture, and the whole shell (except near the point) perfectly smooth: the beak or channel likewise, which in *M. caffra* is short and nearly straight, is in this lengthened and recurved. The mouth is very narrow (occasioned by the outer lip being thick and slightly inflexed) and smooth within, the terminal volutions slightly plaited, and the base of the shell grooved.

The figures of Knorr and Martini are very bad, and give no correct idea of the shell, except its colour.

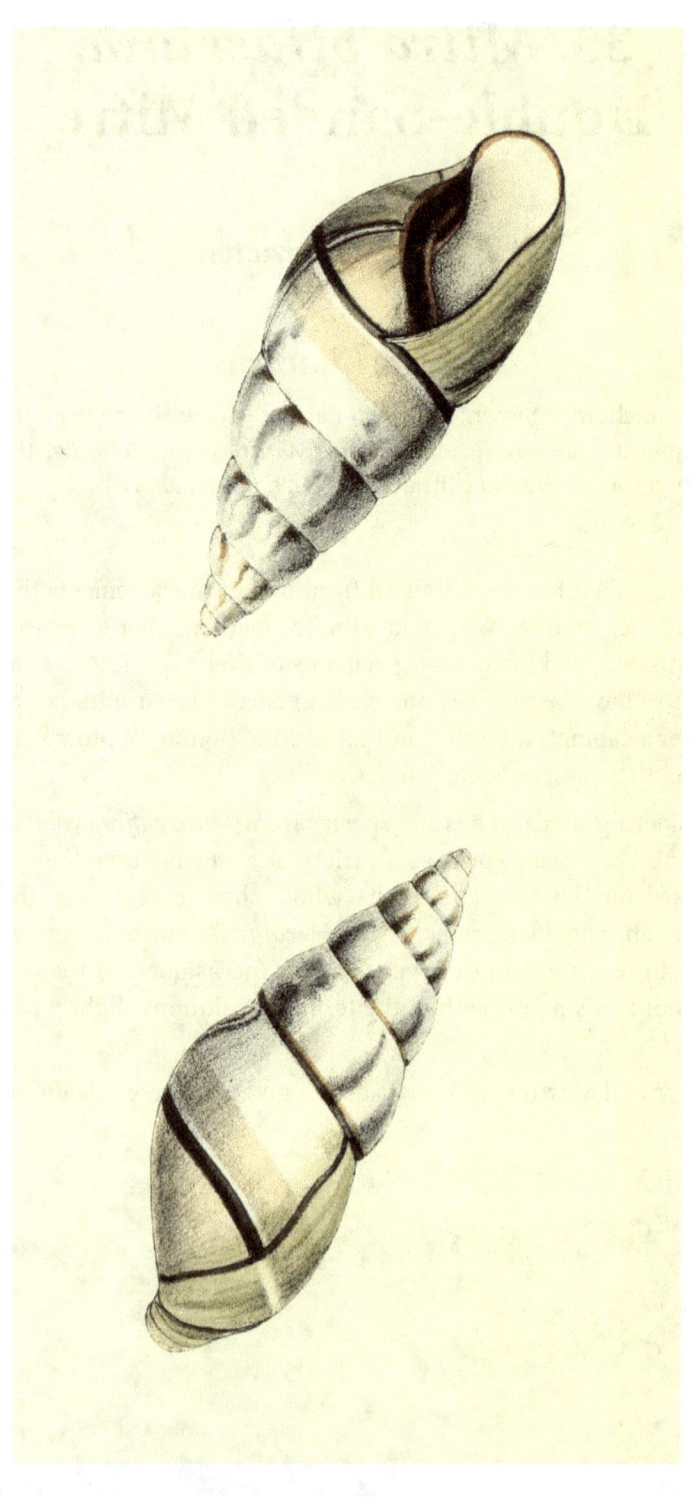

36. *Achatina perversa.*
Reverse Achatina.

Generic Character.

See Plate 30

Specific Character.

Aperture reversed: spire lengthened, of seven volutions, the apex truncated, whiteish with clouded cinereous stripes; central band on the basal volution, pillar, and margin of the outer lip chesnut; mouth within white.

Reverse shells, or such whose mouth when viewed in front is on the left side, are generally held in much estimation by collectors. This deviation from the usual form of shells is sometimes accidental, as in our common garden Snail and several others; while in some species it appears a constant, and therefore a specific distinction. Such I apprehend is the case with the shell now figured, a rare and very elegant species, apparently not noticed by any writer; two or three existing in the British Museum and one in my father's cabinet are all the specimens I have hitherto seen. The latter (here figured) came from Bahia in South America. The whole shell is very finely marked with longitudinal striæ, and the colouring better seen than described: the buff tinge at the base is occasioned by the remaining epidermis.

This shell belongs to the second division of the genus *Achatina* as mentioned at Plate 30 having the aperture much shorter than the spire and the base nearly entire. *Bulla virginea* of Linn. seems to connect the two divisions, having the lengthened spire of one and the truncated base of the other.

37. *Procnias cucullata.*
Hooded Berry-eater.

Generic Character.

See Plate 21

Specific Character.

Head, neck and fore-part of the breast hooded with black; back brown, wings and tail black; tip of the wing-covers, sides of the breast and body beneath yellow; head subcrested.

I am indebted for this new bird to Miss E. Yeates, who received it with a few others from some part of Brazil: it seems to connect the genera of *Ampelis* and *Procnias,* having the bill much less dilated at the base than any of the latter; it however has a close similitude to *Procnias melanocephalus* (Plate 25), which seems further removed from the true Chatterers.

Total length eight inches and three quarters. Bill in extreme length near an inch; the colour dark cinereous; the base furnished with bristles something resembling the Chatterers: the opening of the nostrils large, round, terminal, and nearly naked; the feathers on the crown lengthened; the whole head, neck, and fore-part of the breast black, bordered above by a narrow collar of yellow; back and scapulars brown, rump olive; sides of the breast, inner covers, and under parts uniform yellow; wing-covers black margined with olive, those on the shoulders tipt with brown, the rest with yellow; quills and tail black margined with olive. Wings four inches and three-quarters long, the first quill very short, the third longer than the second. Tail four inches long.

38. *Picus bicolor.*
Black and White Woodpecker.

Generic Character.

See Plate 14.

Specific Character.

White: neck above, back, wings, and line from the ears to the nape, black; tail-feathers black, with their base and spots on the inner margin white.

The simplicity of colouring in the plumage of this bird will easily distinguish it from among the numerous and intricate species already known of this family. It is one of the new birds the recent investigations of Brazilian zoology have added to our museums. The individual here figured was sent me from the district of Minas Geraies.

Total length eleven inches and a half. Bill from the upper base to the tip one inch one line, and from the gape one inch four-tenths; the colour blueish-black; the upper mandible above sharply carinated and slightly curved; orbits (in the dead bird) yellowish-white; the whole of the head and nape, sides of the neck, rump and tail-covers, and all the under plumage pure white, with a tinge of yellow down the middle of the belly: a narrow black line commences at the ears, and is carried down on each side, joining the black of the upper neck; the wings and remaining upper plumage are of a uniform dark sooty black; the tips of the quills much paler and brownish. Wings six inches and a half long; the inside covers black. Tail four inches, and black banded with white at the extreme base; the two outer feathers on each side with alternate black and white bands on the inner web their whole length; feet and claws dirty-greenish: this was a female.

39. *Hesperia Itea.*

Generic Character.

See Plate 28.

Specific Character.

H. (Div. 2.) Wings above blackish-brown, beneath paler, base fulvous. Anterior with a three-cleft yellow spot. Posterior beneath with a fulvous outer margin and longitudinal line. Thighs rufous.

The descriptions left by Fabricius of this as well as many other extensive families of *Lepidoptera*, are in general so vague and short, that unless a figure is quoted to elucidate them, it becomes totally impossible to ascertain the precise species intended. Such is the case with the present insect, which will not agree with any described by Fabricius, or figured by Cramer.

During my travels in Brazil I never met with this species, but am indebted to my liberal friend Dr. Langdorff, Russian Consul-general at Rio de Janeiro, for the specimens I possess, as well as a number of other rare and fine insects of this family, which were then not in my own collection.

On each side of the palpi adjoining the eye are two yellowish round dots, and another behind: the posterior wings above have a narrow whitish margin, the colour beneath much paler; but the nerves on this, as well as at the tips of the anterior wings, are blackish-brown; the legs at the base and the tarsi are black.

This is a male insect; the other sex I have not seen.

40. *Hesperia Cynisca.*

Generic Character.

See Plate 28.

Specific Character.

Hesp. (Div. 2.) Wings blackish-brown; anterior above with a three-cleft yellow band, which in the female is white; posterior beneath immaculate, chesnut-brown, margined externally with yellow.

The different sexes of this insect will appear so strikingly dissimilar to those who are familiarised only with the nice distinctions that separate the species of European *Lepidoptera*, that this affinity by such may be doubted; nevertheless, observations in their native country, and the close examination of several specimens, will we are persuaded confirm the fact.

The male insect is distinguished (like all the *Hesperidæ*) by having the eyes considerably larger, and the anterior wings more narrowed than in the other sex: in this species the bands on their wings assume the form of three yellowish spots, adjoining which, on the inner side, is a semi-lunular villous mark, an almost constant indication (where it exists) of this sex. The straw-coloured border beneath the posterior wings is narrower and darker than in the female; but in both it forms a slender marginal fringe on the upper surface. Legs deep rufous; antennæ black; the club beneath and lunule round the eye straw-coloured.

Inhabits South Brazil, but is not common.

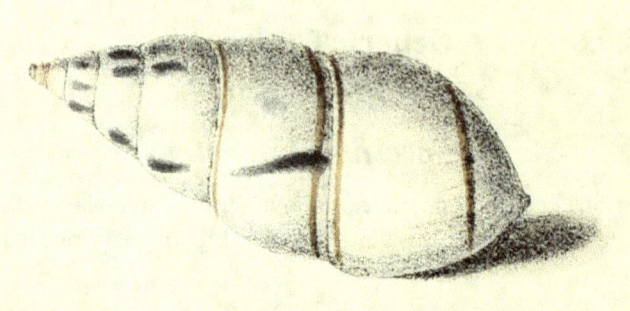

41. *Achatina pallida.*
Pale Achatina.

Generic Character

See Plate 21

Specific Character.

A. Shell cinereous-white, with two narrow brown bands, spire elongated, straight; volutions seven, slightly ventricose, inner lip rosy, base of the columella straight, entire, aperture ovate-oblong.

Obs. another specimen of A. pallida quite agreeing with this, is in Mr. Dubois' cabinet.

The species of this and one or two other genera of land-shells are subject to such variability in their colouring, that it becomes extremely difficult to ascertain which are species and which varieties. The shell now figured might, on a cursory glance, very well pass for one of the Protean varieties of the Linnæan *Bulla fasciata*; but a comparison with that shell will at once point out the strong specific difference that exists between them in the formation of the mouth. In this, the lower half of the inner lip, or more properly the pillar, is nearly straight; the base entire, or without any notch or truncated appearance: whereas in the true *A. fasciata*, the inner lip at the base is very much curved inward, and notched before it joins the outer lip. The mouth is also short and broad: whereas in this it is much more oblong, and the base round. Other more obvious characters exist in the form of the whorls, spire, and more particularly in the colour, of these two shells; but these are in comparison of minor importance.

I regret having but one example of this shell, as it prevents me from tracing how far the characters here detailed hold good in other specimens. They are such, however, as, I think, fully to justify the propriety of considering it a species.

Its locality is unknown.

I have little doubt more than one species exist among the supposed varieties of the true *Bulla fasciata* of Linn., which I take to be the shell figured by Lister.

42. *Oliva Braziliana.* Brazilian Olive

Generic Character.

Shell cylindrical, polished, spire conic acuminated, very short; outer lip simple, inner lip thickened, tumid, columella with numerous slender plaits, aperture at the base truncatedly emarginate.

Generic Type *Voluta Porphyrea* Lin.

Specific Character.

Shell coniform, broad; aperture effuse, tumid callosity on the inner lip large, and spreading over the spire. [Oliva Braziliensis. *Martini p.* 130, *tab.* 147 & 8, 1367 & 8.] [Oliva Braziliana. *Lamarck.*] [Voluta pinguis. *Dill.* 516. 36.]

No family of shells possess characters more strikingly obvious to common observers than the Olives; and yet, although in our English terminology no one would ever think of calling them *Volutes*, we still shrink from giving them that distinguishing appellation in Latin which we every day use and acknowledge in our own language. The strict followers of Linnæus, by thus rejecting generic distinctions, which at once convey a definite idea of form and structure, contribute to render systematic arrangement less expressive of ideas than the common nomenclature of our sale catalogues: a striking proof of the pertinacity with which we cherish those particular doctrines we first imbibed, although an unbiassed reasoning and an attentive observance of nature would convince us of their fallacy.

The great Linnæus, at the time he formed that system which laid the foundation of systematic nomenclature, had not the materials for gathering and combining those natural genera which the immense discoveries made since his death have given us a knowledge of. He accordingly arranged those few shells known to him, in large, and for the most part natural, groups. That of *Voluta* I consider as one of these last (excepting the first division); but the great accession of species now known, and which is still increasing, has long ago induced the principal Continental writers to divide this very extensive family into the following genera: *Marginella* (Date shells), *Oliva* (Olives), *Mitra* (Mitres), *Turbinellus* (Turnip shells), *Voluta* (Volutes), ...; all possessing not only clear but natural characters; inasmuch as, by such an arrangement, those interesting links and ramifications that connect this family with the *Bullæ, Cones, Cowries, Murices,* and other genera, can be traced; and which perhaps affords the most fascinating and intellectual source of contemplation and study the science can bestow.

The peculiarity of this species will distinguish it among this numerous and intricate family. The basal suture is deeply channeled; those on the spire covered by the polished callosity which spreads from the inner lip.

Mr. Dillwyn has adopted the unpublished name of Solander, although the shell had long ago been described and named by Martini and Lamarck. I consider this as contrary to that principle of nomenclature which awards a preference to priority of publication; and I have therefore restored the name of those authors who have this undoubted claim. Mr. Dillwyn's description is very clear and good.

I cannot learn from what particular part of Brazil this species has been received.

43. *Melliphaga auricomis.*
Yellow-tufted Honeysucker.

Generic Character.

Bill moderate, generally somewhat longer than the head, slender, curved, pointed and acuminated, the base higher than broad, the sides compressed, the top carinated; upper mandible notched at the tip, the under mandible laterally compressed. Nostrils concave, near half the length of the bill, covered by a membrane, opening by a long slit midway between the gape and tip. Tongue long, extensible, terminated by cartilaginous fibres. Feet simple; outer fore-toe connected; hind-toe very strong.

Obs. Tail-feathers twelve, first and second quills spurious; margin of the bill sometimes minutely toothed. Generic Type New Holland Creeper Lath., &c.

Specific Character.

Olive-brown: crown of the head and body beneath yellowish; temples and ear-feathers black; throat and lengthened feathers behind the ears yellow. [Yellow-tufted Flycatcher. *Lath. Suppl.* 2. 215. *no.* 4. *Gen. Zool.* 10. 2. 354.]

The Yellow-tufted Honeysucker, although described by Latham, has hitherto remained unfigured; and I therefore select it as an excellent example of a tribe of birds which I think are peculiar to Australasia, and which seem to hold the same situation among the birds of that vast country as the Humming-birds occupy in South America, and the Sun-birds (Cinnyris, Cuvier) in Africa and India; all of which more or less derive their sustenance from the nectar of flowers, and which they extract on the wing by means of their long tubular tongues.

It is singular, that while our first ornithological writers were distributing the numerous species of these birds in their systems, under such of the Linnæan genera as they thought most adapted for their reception, a naturalist of a remote colony should be the first who, by creating a new genus, brought them all into their proper situation in systematic arrangement; one of the many proofs that Nature, and Nature only, is to be studied; and that no system, however ingenious or however applauded, can be considered as infallible.

By an error (no doubt of the press) in the specific character of this bird in Latham's Index, the eye stripe is called white, though in the description it is termed black. Mr. Stephens has copied this error into "General Zoology;" and his description of this bird, as well as number-less others, seems merely an abridgement or alteration of Latham's; a practice highly detrimental to science; for, when an original description cannot be obtained, it is much better, and safer, to copy without disguise that of another.

How far all the birds included by Temminck in this genus really belong to it, admits of very great doubt; I have therefore constructed the generic character from those birds of New Holland only which Lewin, who founded the genus, must have had before him.

Total length seven inches and a half; bill seven-tenths, the frontal feathers advancing half its length to the nostrils; those of the ears are lengthened, but the yellow tuft behind them is much more so; the feathers of the chin are small, thick-set, and ending in fine setaceous hairs curved outwards; the breast and body pale brownish-yellow. Quills and tail dark-brown, margined with deep-yellowish; the two lateral tail-feathers tipt with dirty white; plumage above olive-brown; front and crown of the head dark brownish-yellow; bill black; legs brownish, inner-toe very deeply cleft. Tail, from the rump, three inches and a half long, and slightly rounded.

Latham, who first described this bird, says, "it makes its nest on the extreme pendent branches of low trees or shrubs, and by this means escapes the plunder of smaller quadrupeds." It appears not uncommon in New South Wales.

44. *Pteroglossus sulcatus.*
Grooved-bill Aracari.

Generic Character.

Bill longer than the head, thick, light, curved, thickened at the basal margin, the frontal angle obtuse, the margins serrated. Nostrils nearly vertical, situated on the base of the bill. Tongue long, slender, feathered. Tail elongated, cuneated. Feet scansorial. Generic Type Aracari Toucan Lath. [Illiger. Prod. p. 202.]

Specific Character.

Green Aracari, beneath paler; throat whitish, round the orbits blue; bill with two lateral longitudinal grooves. [P. sulcatus. *Swainson, in Journal of Royal Institution, vol. 9. p. 267.*]

All those species of the Linnæan Toucans having a long wedge-shaped tail, and the nostrils passing through the upper part of the bill, are comprehended by Illiger and other continental writers under this genus. They have been called by the French Aracari; which name I have retained as an English generic distinction. They inhabit the same country and situations as the real Toucans, which are distinguished by having a short, broad, and even tail, and the nostrils placed behind the bill.

A fine example of this very rare bird I first met with in the small collection sent to my excellent friend, E. Falkener, Esq. from the Spanish Main. I have since noticed another which was in Mr. Bullock's museum, and is now in the possession of Lord Stanley: these are the only two specimens known.

This bird was first described by me in the Journal of the Royal Institution near a year ago. When Professor Temminck was in England, I showed him the manuscript description and drawing which I had then made: he assured me he had never seen the bird before, otherwise than in Bullock's museum. A short time after, my account of it was published. I observe, however, that in the new edition of his Manuel he gives this name to a new bird of his own: no description however follows, and it is therefore impossible to say if the Professor intends it for this identical species.

We must postpone any further observations on this family, and conclude by giving the original description above alluded to.

Total length twelve inches, of which the bill in extreme length measures three. It is much curved, and more attenuated than any of the Aracaris, being thickest at the base; from which it narrows to a sharp point at the tip. The upper part is convex, and somewhat thickened; the sides are compressed, and the upper mandible has two broad slightly indented grooves on each side: the base has a few transverse wrinkles, and the serratures deep and unequal. The lower mandible half the depth of the upper, the sides concave, and the teeth less. The colour (in the dried bird) black; the base of the lower and the upper half of the superior mandible rufous, the base with a whitish marginal line. The nostrils are more lateral than usual, being placed in a line with the eye; the orbits naked and reddish brown, the feathers encircling which (particularly beneath the eye) are vivid cerulean blue. The whole upper plumage is parrot green, paler beneath, with a gloss of golden yellow on the cheeks and sides: throat dusky white. Wings short, five inches long, and rounded; inner shafts of the quills black, margined with whiteish. Tail cuneated, green, four inches and a half long, the four middle feathers equal. Legs dusky black.

45. *Ramphastos carinatus.*
Sharp-billed Toucan.

Generic Character.

Bill very large, longer than the head, thick, light, curved, and thickened at the basal margin; the frontal angle transversely sub-truncated, margins serrated. Nostrils vertical, behind the base of the bill. Tongue slender, long, and feathered. Tail short, even. Feet scansorial. Generic Type Red-billed Toucan Lath. [Illiger. Prod, p. 212.]

Specific Character.

Black; throat yellow; pectoral bar and under tail covers red; bill green, tip red; upper mandible carinated and yellow above, the sides with an orange spot; lower mandible varied with blue. [Yellow-breasted Toucan. *Edwards, pl.* 329.] [Ramphastos Tucanus. Yellow-breasted Toucan. *Gen. Zool.* 8, 362, (*excluding the Synonyms.*)]

No tribe of Birds appear so void of that symmetry of form that in general pervades the feathered creation, as the Toucans and Aracaris in the new, and the Hornbills in the old continent. A question naturally arises, why the bills of these birds should be so monstrously out of proportion, and what possible use they can be applied to. The elucidation of these questions is highly interesting, and calls for the most accurate observations to be made in their native regions. It will be sufficient for the present, however, to point out, with regard to the Linnæan Toucans, that the accurate observations and anatomical knowledge of my valued friend Dr. Traill, F.R.S.E., of Liverpool, have clearly proved that an immense number of nerves and fibres fill the cavity of these bills, all connected with the organs of smelling, which are in the highest state of development. A short notice on this subject will be found in the Linnæan Transactions; but as my learned friend is pursuing his inquiries further on the subject, I shall for the present confine my remarks to the individual here illustrated, observing that no birds are so little understood, even in regard to the species, as these.

The indefatigable Edwards appears the first who noticed this bird. His description, though in the quaint style of the day, is clear and comprehensive; and his figure strengthens it, both being made from the living bird. Yet Dr. Latham has quite overlooked it as a variety of another species; and Dr. Shaw, although he copies Edwards's account, gives references which belong to other birds. It is not in the costly work of Le Vaillant, and indeed seems (from its excessive rarity) to have escaped the notice of all modern ornithologists. The perfect bill of the bird is, however, in my possession, minutely agreeing with Edwards's account; and also an original sketch in oil of another individual, by an unknown artist, with a note stating it was done from the life at Exeter 'Change. All these testimonies put the existence of the bird beyond any doubt.

Having seen only the bill, which is well described by Edwards, I shall close this article with such part of his description as appears necessary.

"The bill is very large, compressed sideways, having a sharp ridge along the upper part; the upper mandible is green, with a long triangular spot of yellow colour on each side, and the ridge on the upper part yellow; the lower mandible is blue, with a shade of green in the middle, the point is red, it hath about five faint dusky bars, which cross the joinings of the two mandibles. The iris of the eye is a fair green colour; round the eye is a broad space of naked skin of a violet colour: the throat and breast are of a bright yellow, below which is a bar of scarlet feathers; the covert feathers of the tail are white above, beneath of a bright red; the legs and feet are all of a blue or violet colour." Edwards says it was brought from Jamaica, but doubts its being rather a native of the continent: he says they are very rarely brought home alive.

The bill is full six inches long, and the whole figure on the same scale, both in this and in Edwards.

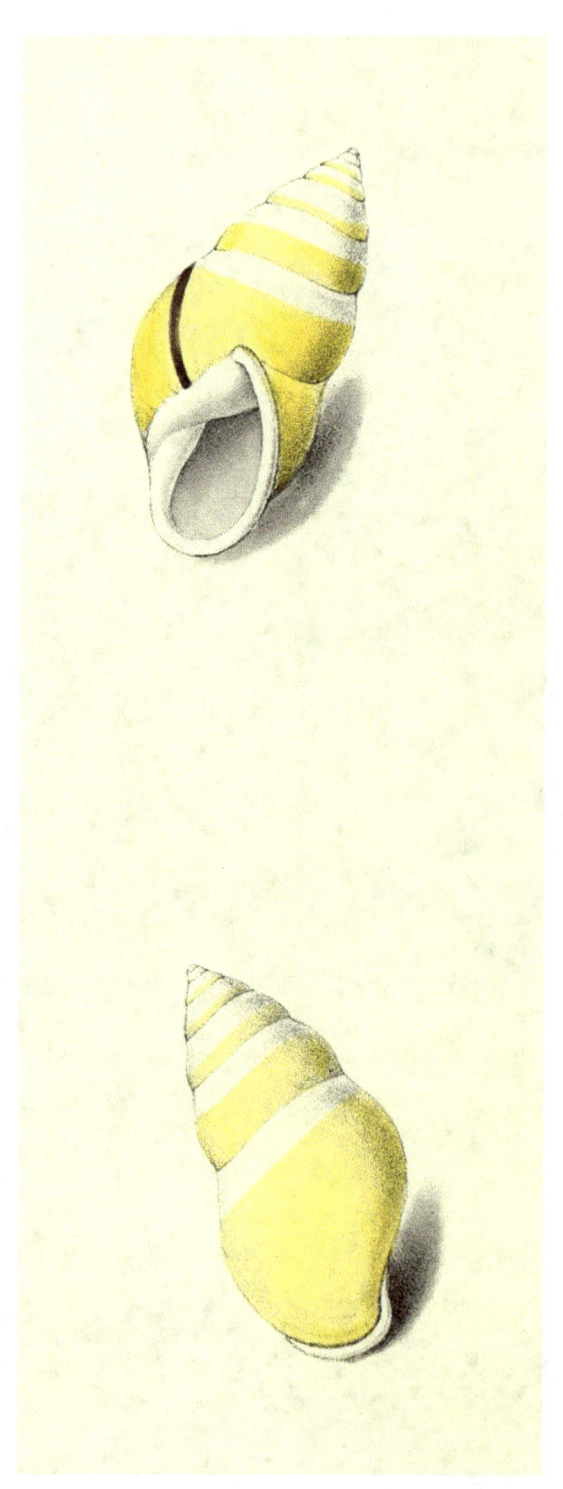

46. *Bulimus citrinus.*
Citron Bulimus.

Generic Character.

See Plate 4.

Specific Character.

Shell obovate; spire conic, slightly thickened in the middle, longer than the aperture, and of six volutions depressed on the suture; outer-lip slightly contracted at the base; umbilicus nearly closed. [Bulimus citrinus, *var.* B. *Bruguiere Encycl. Meth.* 314. *no.* 27.] [*Martini* 9. *tab.* 110. *fig.* 930.]

This variable species is perhaps the most beautiful and delicate in its colouring of all the terrestrial snails; yet, although figured by several of the older writers, so little justice has been done it, that we make no apology for introducing it into the present work, both on this account, and for the purpose of giving such a discriminative specific character as may lead to the inquiry, how far all the numerous varieties mentioned by authors really belong to this species or not. As far as my own observation goes, I have found that the thickened spire, the depression of the whorls on the suture, and the narrowness or contraction of the mouth at the base, afford the only constant characters; for, in regard to colour and the situation of the mouth, both appear subject to great variation, the latter being as often reversed as regular. Martini's is the only figure that can be safely quoted for this variety.

I am indebted to Mrs. Bolton, of Storr's-hall, Windermere, for the loan of this and several other rare shells: it formerly belonged to Mr. Jennings, and appears an old shell, being heavy in proportion, the umbilicus thickly closed up, and the outer-lip very thick. Another I have seen at Mrs. Mawe's, and one is in the British Museum: but the finest specimen in colour and preservation is in the possession of my friend W. J. Broderip, Esq., of Lincoln's-Inn: from this it seems the spiral whorls are finely and delicately marked by transverse elevated striæ, while those on the basal volution are striated transversely, though in a less regular manner.

Bruguiere mentions that this species is generally found in the South American islands, Cayenne, and Guiana.

Mr. Dillwyn has given the new name of *aurea* to this shell, in addition to the five others under which different authors have described it. Such changing of names and multiplication of synonyms, without strong reasons, are very objectionable. I have retained that of *Bruguiere*, as being the only author who has placed it in its proper genus.

47. *Bulimus citrinus, var.*
Reverse, banded Citron

Generic Character.

See Plate 4.

Specific Character.

See Plate 46. [*Martini, vol.* 9. *tab.* 934 & 5. *Knorr,* 4. *tab.* 28. *fig.* 4, 5. (bad.)] [Bulimus citrinus, *var.* B. *Bruguiere,* 314. 27.]

A fine pair of this beautiful and rare variety is in the collection of Mr. C. Dubois, to whom I am indebted on this and many other occasions, for the facilities he has afforded me in prosecuting the present work: one of these is now figured; it differs in no respect from that in the last plate, except in being reversed and having the umbilicus not so completely closed; a character which, perhaps, exists only in very old shells. The other specimen is also reversed and banded, though in a different manner.

Golden Bulimus.—upper and lower figures.

Specific Character.

Shell obovate; spire conic, of five convex volutions; suture simple; umbilicus open.

Obs. Bulimus *aureus*, in Mr. Spurrett's valuable cabinet is another specimen of this shell minutely agreeing with that here described.

H aving seen but a single specimen of this shell, I have placed it as a distinct species, not without some doubts, and principally for the purpose of calling the attention of conchologists to a more rigid examination of the specific distinctions of this family (unconnected with colour) than has heretofore been done. The regular convexity of the whorls, not in any degree compressed at the suture, the want of that thickened appearance on the spire, and of the contraction at the base of the mouth (all which characters I have found in the varieties of *B. citrinus* to be constant), afford a specific distinction which future observations must confirm or annul. Bruguiere notices a variety of *B. citrinus* which is entirely yellow, a most beautiful specimen of which is in the British Museum, and which possesses (as well as the excellent figure of Gualtieri) all the specific characters I have given to *B. citrinus*, but not of the present shell. Lister's figure, on the contrary, is rude, though very characteristic of this; and Martini's representation, here cited, also appears the same.

Mr. Dubois, in whose collection this specimen exists, is unacquainted with its locality.

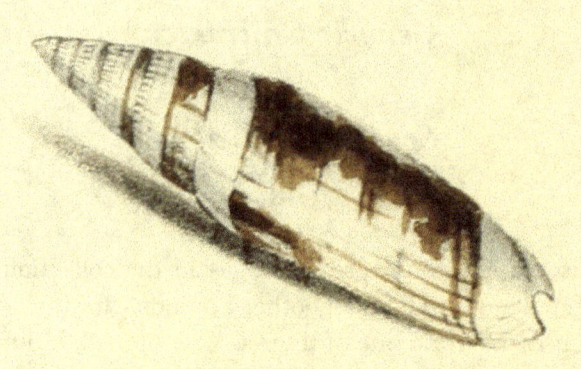

48. *Mitra casta.*
Chesnut-banded Mitre.

Generic Character.

See Plate 23.

Specific Character.

Shell white, smooth, olive-formed, spire shorter than the aperture, the volutions finely reticulated above, the lower half with the epidermis forming a chesnut band which is central and broad on the basal whorl. [Voluta casta. *Chemnitz* 10, *p.* 138, *vig.* 20 C D.—*figura mala.*] [*Martyn Univ. Conch.* i. *Tab.* 20.] [*Dillwyn Catalogue, vol.* i. *p.* 554, *no.* 127.]

All the writers I have been able to consult, uniformly describe this species as having a coloured band on the white ground of the shell. In the Banksian cabinet are two fine specimens, and which have enabled me to ascertain that this brown band is nothing more than an epidermis, or external coating, with which the shell is only partially covered—a circumstance of very rare occurrence; and which, being removed, proves the real colour of the shell to be of a uniform polished white. This, together with its excessive rarity, and the opportunity of giving original figures, has induced me to include it in this work, although it exists both in those of Martini and Martyn above quoted. I have neither seen nor heard of specimens being in any other collection, besides the two above noticed; and which no doubt were collected by their late lamented possessor on some of the South Sea islands. A striking affinity exists between this and *M. zonata* figured at the third plate of this work.

Olive-shaped Mitre.

M. Shell olive-shaped, smooth, polished, spire very short, longitudinally wrinkled, with a central transverse stria, pillar four-plaited.

I introduce the description of this diminutive and undescribed shell from its affinity with the last, and as forming a most interesting transition from the Mitres to the Olives: agreeing with the former in the structure of the pillar and the sculptured spire, and with the latter in its general form and *prima facie* appearance. Its perfect resemblance, in fact, to a small olive, may have occasioned its being hitherto overlooked. The spire is slightly wrinkled and striated; the teeth on the pillar very near each other, slender, and four in number. The colour pale yellowish; the mouth darker, and the tip and base purple. The whole shell is scarcely half an inch long.

It was received from the South Seas.

42

49. *Oxyrhynchus cristatus.*
Crested Sharpbill.

Generic Character.

Bill short, very straight, base trigonal, beyond attenuated to a very fine point; upper mandible above rounded, both entire. Nostrils basal, naked, partially covered by a membrane; aperture linear, near the margin of the bill. Feet short, strong, a little longer than the middle toe; anterior toes three, the outer connected, the inner cleft; hind toe strong.

Specific Character.

Above olive-green, beneath yellowish-white, with blackish spots. Head with an incumbent crimson crest; sides of the head and neck with transverse yellowish-white lines.

An elegant and (to the ornithologist) a highly interesting bird, considered with much judgement by Professor Temminck as a new genus, having the perfect bill and habit of the Wryneck, but totally unlike that bird in the position of its toes, which in this are not placed in pairs. The Professor has slightly described it, in the new edition of his *Manuel*, without a *specific*, but under the *generic* name of *Oxyruncus*, the spelling of which must be presumed as an error of the press: no mention, however, is made of the beautiful crimson colour which adorns the crest.

Total length near seven inches. Bill eight-tenths in length from the gape; general colour of the bird olive-green, becoming nearly white on the under part, and on the transverse stripes on each side the neck, front and temples, where there are also obscure bands of black; crown with a concealed crest, which is vivid crimson at the base and blackish at the tips; inner margin of the covers, quills and tail blackish; inner covers yellowish; chin, neck and breast banded with blackish lines, which are broken into spots and stripes beyond.

Inhabits Brazil, but is very rare.

50. *Alcedo Asiatica.*
Asiatic Kingsfisher.

Generic Character.

See Plate 26.

Specific Character.

Head black, transversely banded with mazarine blue, the hinder part crested; ears blueish; chin, throat, and lateral stripe on each side the neck whitish; back shining light-blue; body beneath rufous.

Obs. This bird Dr. Horsfield tells me is his *Alcedo meninting* described in the Linn. Transactions.

The general resemblance between this and the European Kingsfisher may have been the cause why it has remained hitherto unnoticed by ornithologists. It bears, however, on closer inspection, a strong and peculiar distinction in the crest at the back of the head, in being much smaller in size, and especially as inhabiting the hottest parts of India; while our own braves the cold of a Siberian winter.

Total length six inches, of which the bill from the angle of the mouth to the tip occupies one inch and three-quarters, and is black, with the under mandible paler; the ears and the upper part of the head and neck are blueish-black, transversely banded with somewhat crescent-shaped narrow bands of a rich deep blue, which are broken into spots on the crest and ears: from the base of the under mandible is a black stripe richly glossed with blue, and carried down on each side the neck, between which and the upper part is a whitish stripe beginning just behind the ears (this in the European species is rufous). The wing-covers, scapulars and lesser quills are blackish glossed with blue, the two former having a bright spot at the end of each feather; superior and greater quills entirely blackish; down the middle of the back, rump, and tail-covers, light and vivid blue, with a slight tinge of greenish; chin and throat cream-colour; line between the nostrils and eyes, margin of the shoulders, under wing-covers, and all the lower parts of the body, rufous; tail deep and obscure blue; legs red.

My specimen came from some part of India; I have met with others from the same place; and Dr. Horsfield has likewise observed it in Java.

51. *Colias Pyrene.* White African Colias.

Generic Character.

See Plate 5.

Specific Character.

Wings white; anterior with a small, nearly central, oblong, black dot nearest the tip; margin of the posterior wings very entire, beneath all with a brown ocellate spot and undulated fulvous lines: both sexes alike.

Under the head of *Colias Pyranthe*, M. Godart has united the three insects described by Fabricius, as, *Pap. Pyranthe, Nepthe* and *Gnomia*, all bearing in their leading colours a very near resemblance to each other. Yet as this consideration alone appears to have decided this ingenious author in uniting them, without apparently noticing the nicer but more important characters of form, proportion, and real sexual distinction, as well as geography, I cannot but consider the question still remains doubtful; and although I am not at present prepared to offer an opinion as to the actual affinity between these three insects, I have little or no doubt that the one now figured is a really distinct species from either of the above, which all inhabit various parts of India. This, on the contrary, is from the interior of the Cape of Good Hope, from whence it was brought by Mr. Burchall, among whose insects I have seen about twenty unvarying specimens, but they were all males. I discovered however three or four of both sexes, varying in size, in Mr. Haworth's cabinet, and the perfect similarity in colour of the female with the other sex is very striking: it wants of course the little tuft of hair and opaque spot within the borders of the wings, so generally found in the male *Coliadæ*.

The distinctions of *Colias Pyrene* as a species rest on the areola of the anterior wings being considerably larger in proportion than in the others allied to it, thus making the black dot (which is always placed at the outer extremity of the areola) much nearer the tip than the base: these wings are also more sharply trigonal (in the male), and have only the slightest appearance of a black margin; the hinder wings are also perfectly entire, and not obtusely undulated as in those insects, and the sexes not differing in colour. Like most of the insects of this genus, the ocellate spots beneath vary considerably; sometimes they are silvery, at other times not; the anal valves in the male are short and obtuse, and the wings in the female not so sharply pointed.

52.

52. *Colias Argante.*
Orange Colias.

Generic Character.

See Plate 5.

Specific Character.

C. (Male.) Wings bright-orange, above immaculate; posterior beneath with minute ferruginous dots, and generally two silvery spots. [Papilio Hersilia. Cramer, *pl.* 173. C. D.] [Papilio Hersilia argante. *Fab. Ent. Syst.* iii. *pt.* 1. *p.* 189.] [Colias argante. *Godart in Encycl. Method.* 9. 92. *no.* 11.]

C. (Female.) Wings golden-yellow; anterior above with a central spot and black marginal tip; posterior beneath covered with ferruginous dots and two silvery spots. [Papilio Cipris. *Cramer, pl.* 99. E. F.] [Colias Cnidia. *Godart*, 93. *no.* 14.]

No two insects can present a more striking dissimilarity than the sexes of this species; and it was only after a considerable degree of attention to the subject, in their native climate, that we were at last thoroughly convinced that *Colias argante* and *C. Cnidia* were, without the least remaining doubt, the male and the female of one species. I have had the same opinion communicated to me by my friend Dr. Langsdorff, whose long residence and observations in Brazil render his opinion of no small authority.

As both insects are well known, and their distinctions given in the specific character, it will be only necessary to observe, that the under surfaces of the wings in both sexes vary much both in the density of the minute dots, and short undulated stripes that spread over their surface; and that the silvery spots in some males are strongly marked, and in others quite obsolete: there is, in very fine individuals of this sex, a faint bloom of pink spread on the orange of the upper surface, which heightens the vivid yet chaste beauty of the insect. The females are not so common. I met with them both in northern and southern Brazil, and have seen them in collections from Parà directly under the equinoctial line.

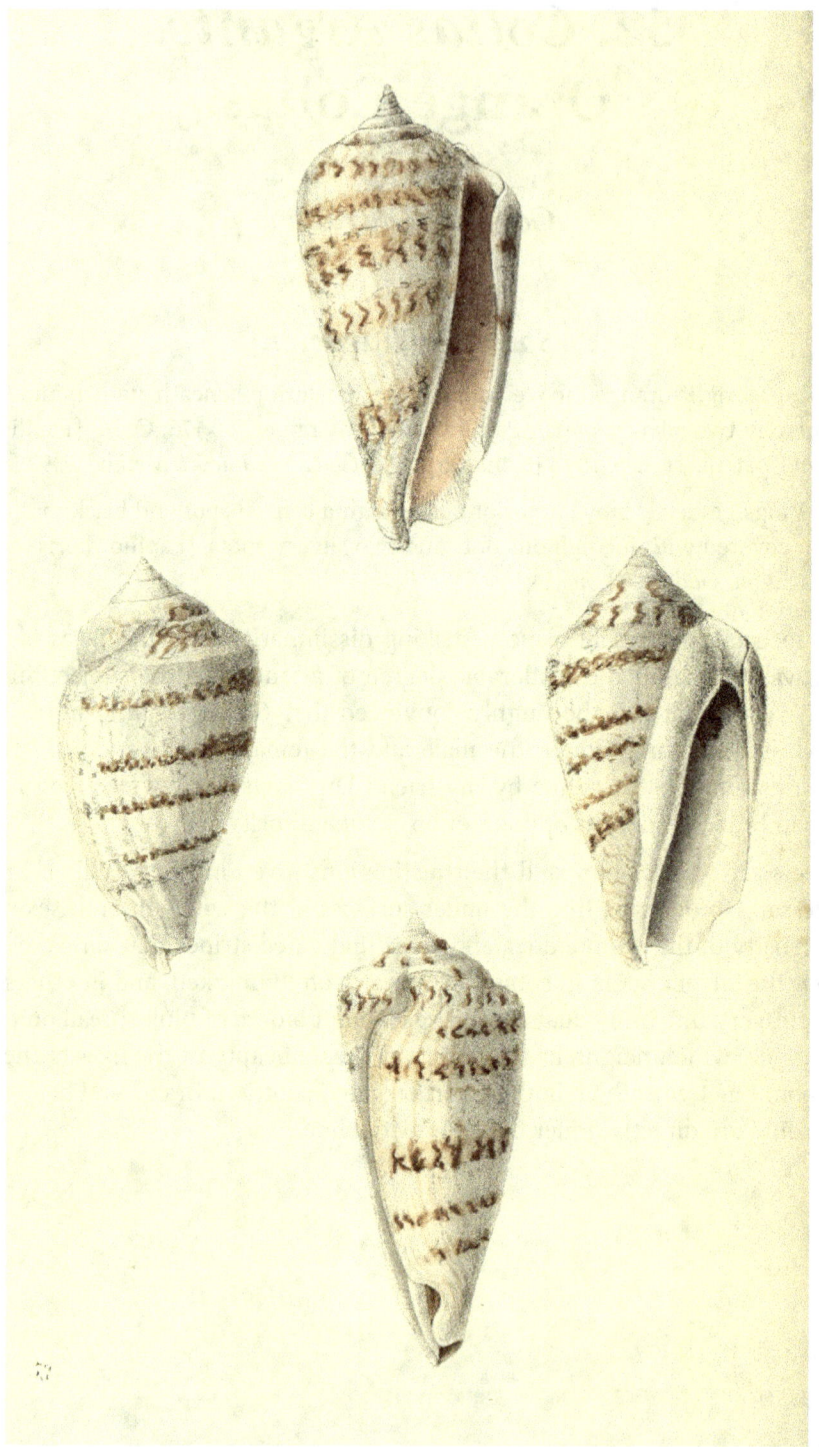

53. *Strombus.*

False Scarlet-mouthed Strombus—Upper and under figures.
Generic Character.

See Plate 10.

Specific Character.

Shell coniform; spire short, depressed at the base, the whorls convex and unequal; outer lip lobed above, and internally striated; inner lip nearly obsolete, white. [*Lister* 850. 5. (bad.). *Gualt.* 31. 1. *Knorr*, vi. *Tab.* 15. 3.] [Strombus luhuanus *Linn. Martini*, x. *Tab.* 157. 1499. 1500.] [Young. Lip above entire, inside smooth, whorls tuberculated. *Lister*, 849. 4. a? *Knorr*, vi. *tab.* 17. 2.]

We introduce this common shell for the purpose of pointing out those characters which induce us to consider it more as a distinct species than as a variety of *S. Luhuanus* of authors; and this consists not so much in the colour of the inner lip, as in the almost total absence of that important part, which this shell invariably exhibits through all its growths: it is therefore, I think, contradictory to the meaning of the word to term that variable which is found to be constant, particularly where the point of distinction rests on a marked difference of *formation* no less than of colour, though both shells are common to the Oriental seas. Minor differences exist, in the mouth of this always being pink, the inner lip white, and the outer lip but slightly lobed (or cut out) above: in *S. Luhuanus* the mouth is deep scarlet, inner lip black and highly polished, and the outer lip deeply notched above; the two former characters, indeed, begin to show themselves at a very early growth of the shell.

Persian Strombus—middle figures.

Shell somewhat coniform, short; spire conic, equal. Outer lip prominent, sinuated above, smooth within; inner lip polished, white.

Allied, but sufficiently distinct from the last; the mouth is always smooth and white. It is a local species: a few received from the Persian Gulf are all I have yet seen, some were young, but no other variation was observable.

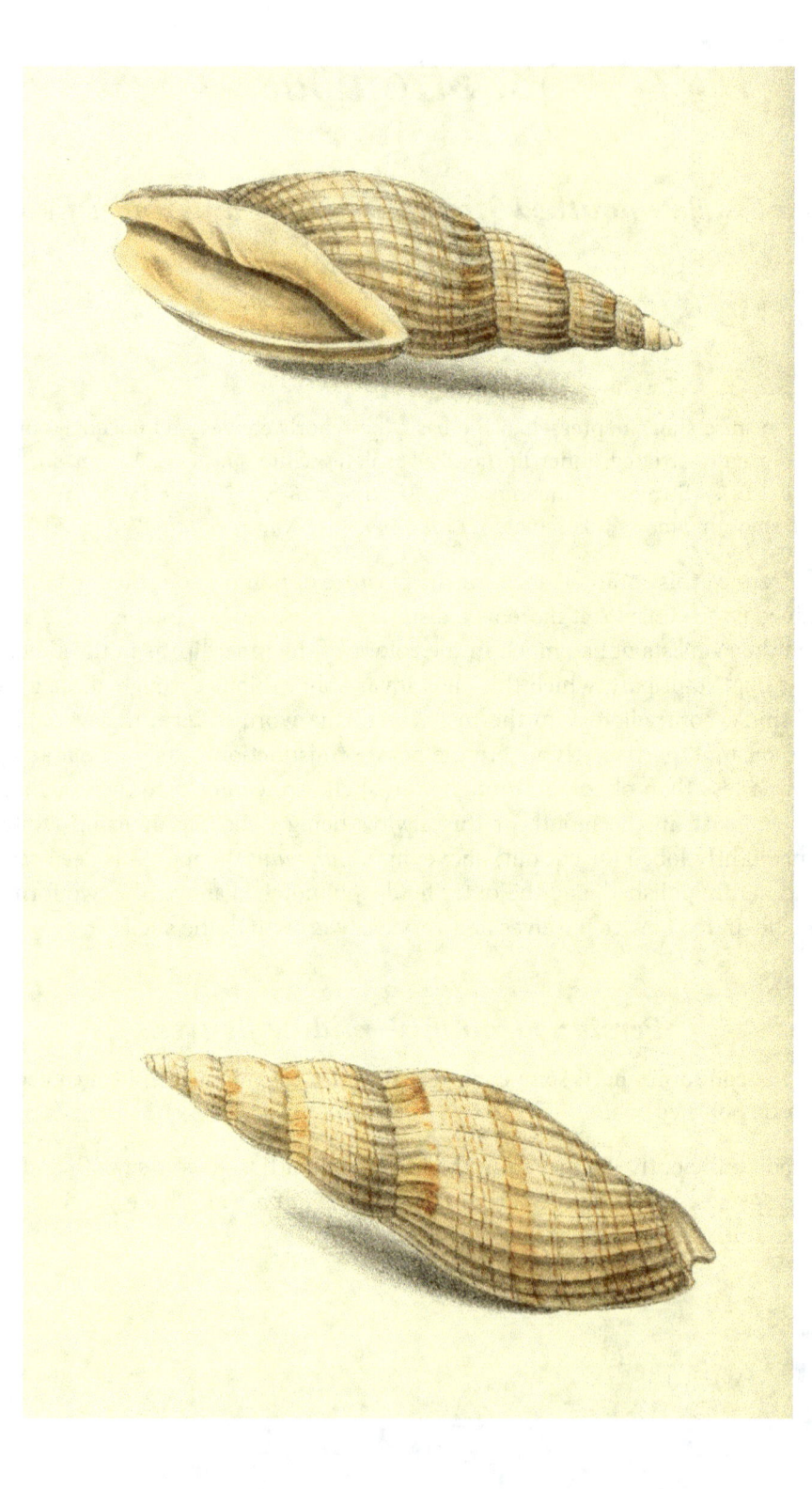

54. *Mitra lyræformis.*
Harp Mitre.

Generic Character.

See Plate 23.

Specific Character.

Shell with regular, carinated, approximating, longitudinal ribs. Pillar striated, three-plaited near the base. Spire somewhat attenuated. Apex slightly papillary.

This beautiful and highly interesting shell has been generally considered *unique* among the collections in this country. It was originally in the possession of the late Mr. Jennings, and, I am informed by Captain Laskey, was on first being received, in a much finer state. Mr. Jennings had it cleaned, and in so doing many of the delicate transverse striæ were partially obliterated, and the sharp ridges on the longitudinal ribs worn down, as indeed was apparent from a drawing Captain L. had made of the shell previous to this unmerciful cleaning. It however still remains a very fine shell, and is now in the cabinet of Mrs. Bolton, of Storr's-hall, to whom I am obliged for the opportunity of now publishing it.

The figure and specific character will sufficiently point out its distinctions. The body-whorl is smooth, but strongly granulated at the base; the spire delicately striated between the ribs; the two last whorls before the apex are close, thick, and somewhat papillary; the apex itself small and sharp. The upper part of the inner lip has some faint obsolete teeth, but the base has three very strong ones.

It connects in the most beautiful manner the two genera of *Mitra* and *Voluta*. Its country is unknown.

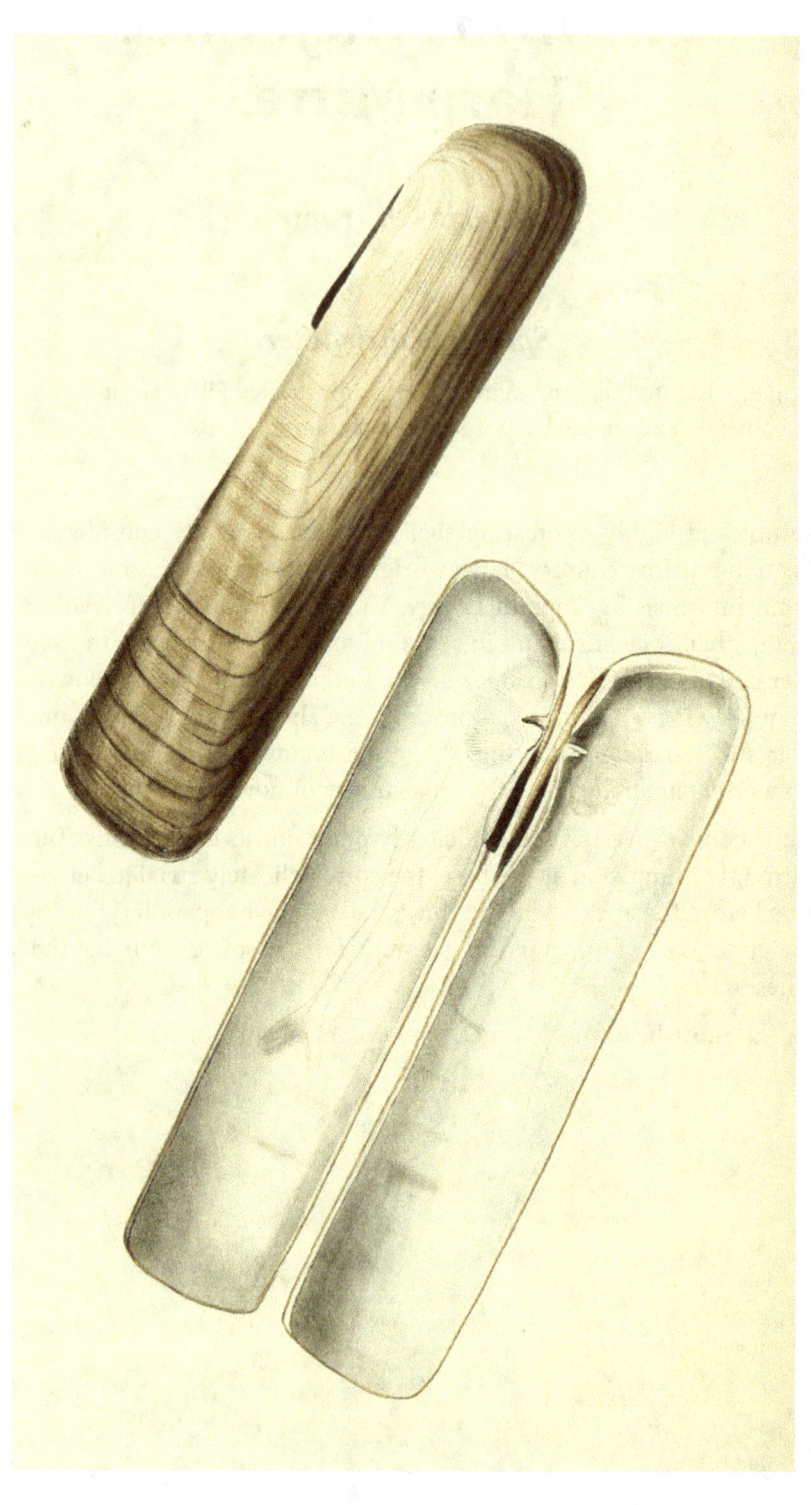

55. *Solen ambiguus.*
Ambiguous Solen.

Generic Character.

Shell bivalve, equivalve, very transversely elongated, open at both ends. Cardinal teeth small, fragile, variable in number, and rarely divaricated. Ligament external. Animal with a sub-cylindrical foot at the anterior end, and at the other a short tube containing two others united together. *Lamarck.* Generic Type *Solen Vagina* Pennant.

Specific Character.

Shell linear, strong, straight, pale, obscurely radiated. Cardinal teeth one in each valve, placed near the anterior extremity. [Solen ambiguus. *Lam. Syst. vol.* iii. *p.* 452. *no.* 7.]

Under the genus *Solen* (vulgarly called Razors or Pods) are comprehended a variety of shells having the common character of both extremities open or gaping when the valves are together, yet differing materially in their form, teeth, and general appearance: some are long, slender and straight; others more or less curved; a few short and oval, or with one end only lengthened. Modern writers have, however, retained nearly all these in the genus as left by Linnæus; and this method for the present is more desirable than that of creating a multiplicity of genera. Dr. Turton, in his very useful Conchological Dictionary, enumerates thirteen species as found on the British coast, including the *Solen Novacula* of Montagu, which the Doctor suspects is not truly a species. The original specimens which Montagu described I have carefully inspected at the British Museum, and have no doubt in my own mind they are in reality no other than *S. Siliqua* with one of the cardinal teeth broken off; a circumstance which, from their fragility, frequently happens, even in opening the recent shell.

Solen ambiguus was first described by Lamarck, who says it is from North America. Two or three specimens are in my possession; but it is a rare species, much thicker, and with larger teeth than any other; the epidermis is pale-brown, and in some parts obliquely lineated.

56. *Ramphastos vitellinus.*
Sulphur-and-white-breasted Toucan.

Generic Character.

See Plate 45.

Specific Character.

Black; throat yellowish-orange; the sides and ears white; pectoral bar and tail-covers red; bill black, with a blue basal belt, the top convex and but slightly curved, the sides thickened. [Le Pig-nancoin. *Vaill. pl.* 7.] [*Var.?* Le Grand Toucan à ventre rouge. *Vaill. pl.* 6.]

The descriptions of Dr. Latham, and the compilations of Dr. Shaw on the various species of Toucans, are so confused, and their synonyms so inaccurate, that it is quite impossible to quote them in reference to this bird; but which I am informed has already been distinguished by the celebrated Illiger as a distinct species, under the name here adopted.

Independent of colour, this differs from *R. Tucanus* in having the bill less curved, the top convex and obscure pink, not flat and blue. The belt at the base is always vivid blue (grey in the dead bird), not, as in *R. Tucanus*, of a rich yellow. This I have never met with in Brazil; the other is common from lat. 8 to 23° S. A drawing from the live bird by the late Sydenham Edwards (obligingly lent me by Lord Stanley) confirms others I have seen as to the colour of the bill, orbits, &c. It varies, however, in that of the throat, breadth of the red band, and in the tail-covers. A specimen I possess being somewhat larger, the breast is nearly white, and the upper tail-covers sulphur. In young birds the white on the sides is tinged with grey. I am inclined to consider the *Grand Toucan à ventre rouge* of Vaillant as a mere variety, having the red pectoral bar very broad.

In general size it is rather larger than the Brazilian Toucan. Our figure is on the exact scale of four-tenths to an inch. Its precise locality I am unacquainted with. We hope to enlarge more on this interesting genus in another publication.

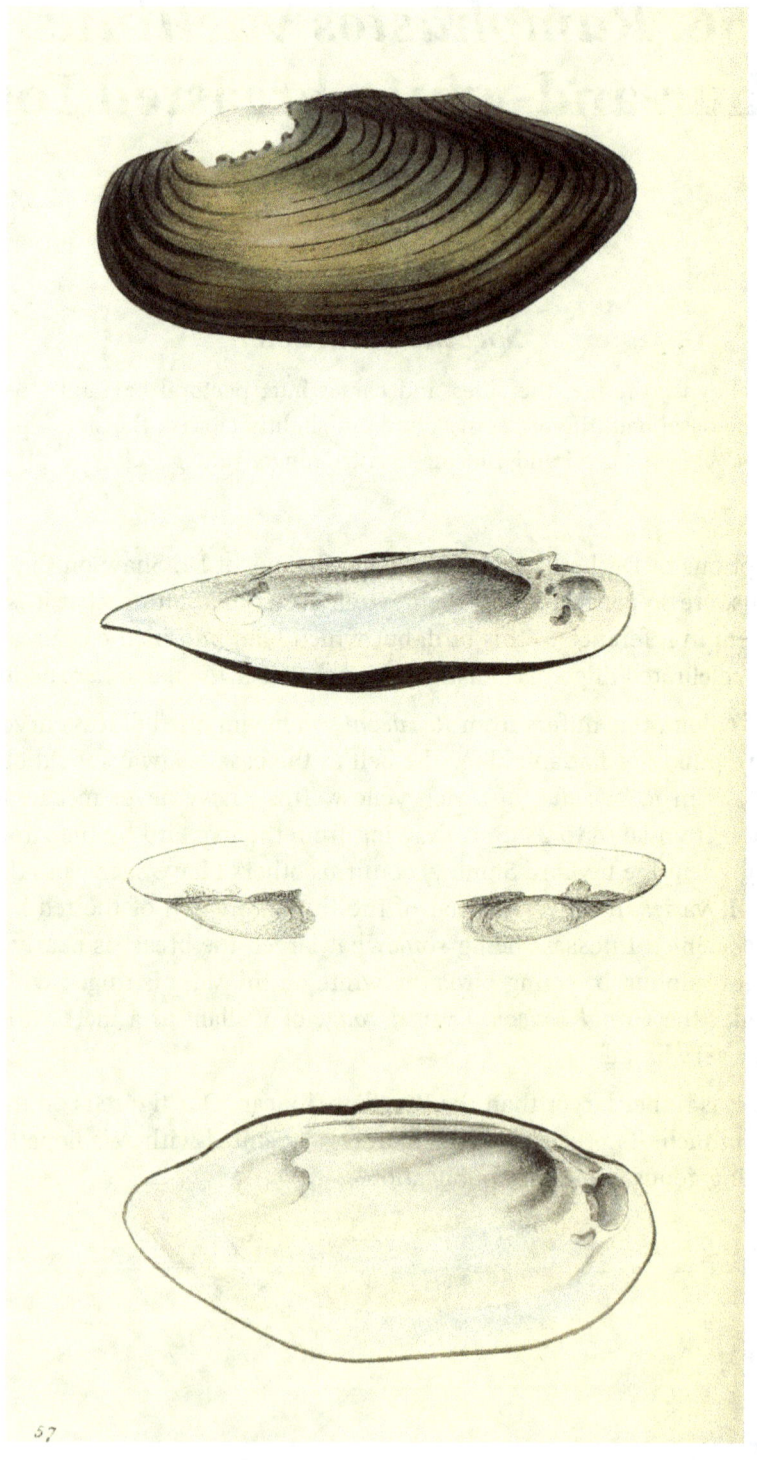

57. *Unio nasutus.*
Rostrated River-Mussel.

Generic Character.

Shell transverse, equivalve, not affixed, the tops decorticated. Posterior muscular depressions double. Cardinal tooth one, short, irregular, simple or double, striated; lateral teeth two, elongated, compressed, and prolonged beneath the corslet. Generic Type *Mya Pictorum*. Linn.

Specific Character.

Unio (Div. 2.). Shell transversely elongated: dorsal margin straight; anterior side angulated, obliquely attenuated, the extremity slightly truncated. [*Lister, tab*. 151. *fig*. 6.] [Unio nasutus. *Say in Encycl. Am. Conch. pl.* iv. *fig*. 1.]

This is one of the most natural genera in the modern systems of conchology, as it includes all fresh-water bivalves having two rough cardinal teeth in one valve and one in the other. The colours of all are more or less dark-brown, sometimes radiated with green; but the specific characters rest on the contour of the shell and the proportion of the teeth.

There can be no doubt this shell is the *Unio nasutus* of Say, who refers to the figure of Lister. The *Unio nasuta* however of Lamarck I apprehend will be found different, as he seems to think; his shell also is purple inside with short thick teeth; indeed so much uncertainty hangs on the shells of this genus, that the species can only be fixed by ample descriptions and very correct figures. The figure is from a shell in the Linnæan Society's cabinet. Mr. Say says it is common in the Delaware River, North America.

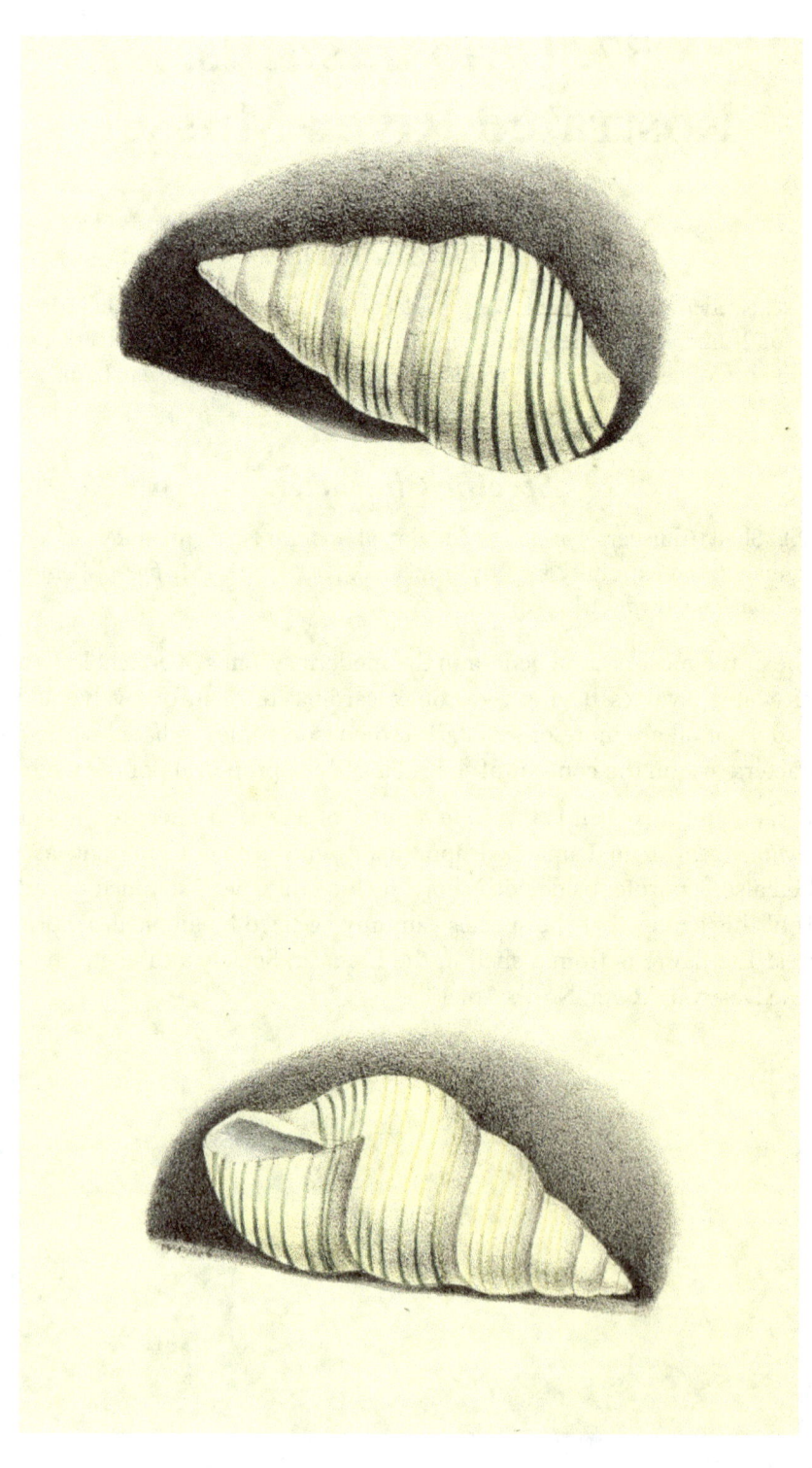

58. *Achatina crenata.*
Green hair-streaked Achatina.

Generic Character.

See Plate 30.

Specific Character.

Shell white, with capillary green bands; spire elongated, sub-attenuated, of six convex volutions; outer lip crenated; base slightly truncate.

A few specimens of this most delicate and beautiful shell were found by my brother, Mr. J. T. Swainson, jun. in the island of Cuba; nor am I aware of its having by any other means reached our cabinets, excepting a distinct variety which occurred in Mrs. Angus's, and is now (together with a young one of the same) in Mr. Dubois's collection. This has, in addition to the green bands on the spire, a row of bead-like cinereous spots at the base of the first and second spiral whorl; and others of a longitudinal square form on those whorls nearest the tip, which, with the inner lip, is slightly tinged with pink. The form of the shell also is shorter; but the general contour, and particularly the crenated mouth, common to both, clearly proves it can be considered only as a distinct variety. The specimen we have figured agrees with all those sent at the same time, in having not the slightest appearance of spots, though in a perfect state of preservation. The little notches on the margin of the lip are always placed at the commencement of each of the green lines; the base of the column is straight, and slightly truncated before it joins the outer lip.

59

59. *Psittacus Barrabandii.*
Red-collared Parakeet.

Generic Character.

See Plate 1.

Specific Character.

Green, fore-part of the head and throat yellow; round the middle of the neck in front an orange-red collar; bill red; spurious wings blueish.

The vast and little known region of New Holland has afforded us some of the most beautiful birds of this superb family, and among which the species now, as we believe, for the first time published, will stand conspicuous. It is from a fine skin in the possession of Mr. Leadbeater, and is named in honour of the late M. Barraband, the first ornithological painter that France or any other country has produced.

The tail is very long, measuring eight inches three quarters; the total length of the bird being near fifteen inches. The green which predominates over the plumage is bright and changeable, having a blueish tinge on the hind head, which is much darker and stronger on the outer margins of the quills and middle of the tail-feathers: the back and scapulars are tinged with an olive-brown; the spurious quills and their protecting covers are greenish-blue, appearing in some lights entirely of the latter colour; the inner margin of the quills and tail, as well as their entire under surface, deep brownish-black; but the tips of the tail-feathers beneath are much paler; the two middle feathers five inches longer than the outermost, and extending near two inches beyond any of the others; their extremities instead of being pointed are rather widened and rounded. Bill red; ears and space between the eye and bill green; fore-part of the head, chin, and half the neck, a clear orange-yellow, which is terminated by a narrow collar of a beautiful orange-red; the remaining under plumage pale-green; inner wing-covers darker. Legs black. The fourth, fifth and sixth quills notched at their tips.

60. *Thyreus Abbottii.*

Generic Character.

(*Familia Sphingidis* Latreille.) Antennæ linear, thickened in the middle, externally ciliated in the male, simple and filiform in the female, gradually ending in an arcuated, obtuse hook. Palpi short, obtuse, alike in both sexes. Wings opaque, angulated. Abdomen thick, bearded.

Specific Character.

T. Wings angulated; anterior testaceous, with lineated brown shades and oblique lines; posterior yellow, with a broad black border.

A lovely insect, which is unfigured, and, as far as we can ascertain, undescribed by any author. It appertains to the Linnæan genus *Sphinx*, which can be viewed (from the immense diversity and great number of the species) only as a family containing many and striking natural genera: in modern arrangement it is most nearly allied to the *Sesiæ* of Fabricius, from which, as it strikingly differs in the formation of the palpi and antennæ, I have separated it.

I have named this insect to commemorate the exertions of Mr. Abbott, well known as having furnished the materials for that beautiful work the Lepidopterous Insects of Georgia, edited by Sir James Edward Smith. And from the unpublished drawings of this zealous collector, the larva and pupa have been figured. Mr. Abbott writes that it is a rare species in Georgia, and feeds on the grape. The female differs not in colour from the male, which is here represented.

61

61. *Tamyris.*

Tamyris Nurscia.
Generic Character.

See Plate 33.

Specific Character.

Wings black; anterior above with a central reddish band, and two white basal dots beneath; posterior beneath grey and cinereous; base black with an obsolete white line; margin black.

The marginal fringe of the lower wings in this species has a few white dots between the nerves, and the upper surface is sprinkled or powdered in the middle with blueish-green atoms; on the under surface of the anterior wings the lower part of the band is orange, the upper bright rufous; and within the black margin of the posterior wings is a large blueish spot, and two or three whitish dots on the sides of the thorax. It seems nearest allied to *Hesp. Celsus* of Fabricius, which is only slightly described from Mr. Jones's unpublished drawings.

Tamyris Laonome.—*lower figure.*
Specific Character.

Wings in both sexes alike, uniform brown, with a common margin of orange; head and tail red.

The under surface of this insect (a female) perfectly resembles the upper: it will approach near to *Hesp. Amiatus* of Fabricius, which no doubt belongs to this genus.

For both these interesting insects, not to be found in Fabricius, I am indebted to the liberality of my friend Professor Klug, Director of the Royal Museum at Berlin: no note accompanied them, I therefore conclude they are undescribed, and probably inhabiting South America.

62. *Psittacus discolor.*
Red-shouldered Parakeet.

Generic Character.

See Plate 1.

Specific Character.

P. Green; front, chin, under wing-covers, and spots in the neck, flanks and scapulars crimson; shoulders dark-red; crown, external wing-covers, and lateral tail-feathers violet-blue; quills blueish-black margined with yellow. [P. discolor. Red-shouldered Parakeet. *White's Voyage, pl. at p.* 263.] [La Perruche Banks. *Le Vaill, pl.* 50.]

This is another of the splendid little Parakeets inhabiting the forests of New Holland; and vivid as the colouring may appear in our figure, it sinks into dullness when compared with the bird itself. Dr. Shaw was the first who described it in White's Voyage to New South Wales, where it is badly represented. It has been since figured by Le Vaillant, probably from a female or imperfect specimen, as the tail is represented by far too short, and the colours not quite agreeing with that in my collection.

Total length eleven inches. The upper plumage bright green, tinged with blue on the sides of the neck, lighter and yellowish beneath; the crown of the head sapphire or violet-blue, with a crimson belt in front, and a large patch of the same round the chin; paler spots of this colour are also in front of the neck, breast, flanks, and under tail-covers; the under wing-covers are deep crimson, as well as the inner shafts of some of the lesser covers outside; the shoulders dark blood-colour; the outer wing-covers deep-blue on the margin of the wings, gradually changing to a vivid blue, which blends with the green. Quills black glossed with violet, margined externally and internally with yellow. Tail near five inches long, the middle feathers dark rufous tipped with blueish; the rest more or less rufous at the base, and shining blue beyond. Bill and legs pale.

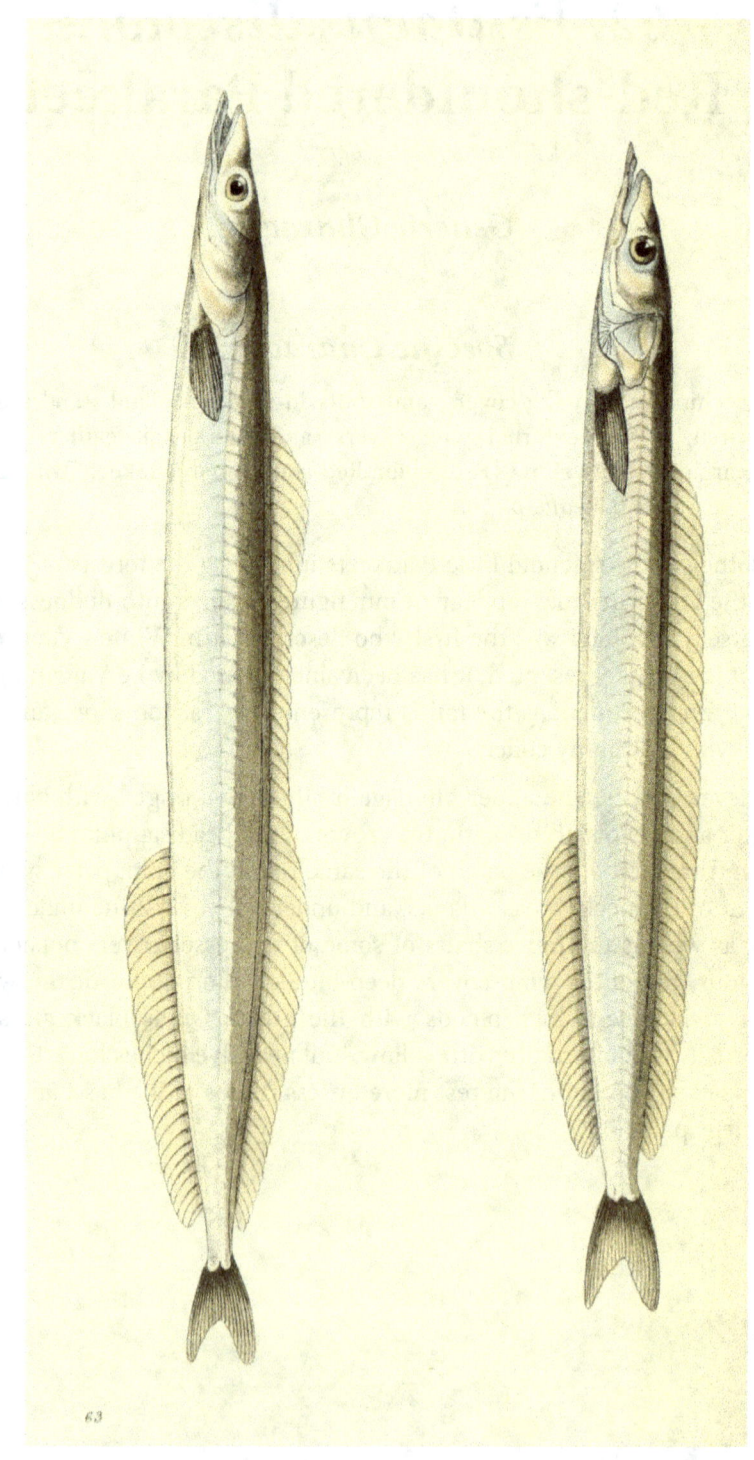

63

63. *Ammodytes Tobianus.*
Sand-Lance.

Generic Character.

Body slender, roundish, many-sided, with minute scales. Upper lip doubled; lower jaw narrow pointed. Gill membrane seven-rayed. Dorsal fin nearly as long as the body, with simple flexible rays. Generic Type *Sand-Lance.* Pennant.

Obs. The *Ammodytes cicerelus* of my friend Professor Rafinesque must be different from *A. siculus*

Sicilian Sand-Lance—left figure.

Dorsal fin sinuated, narrowed in the middle and broadest behind.

Of this genus, hitherto considered as possessing only a unique example, we were fortunate in discovering while in Sicily the new species now figured, and which early in the year visit the coasts near Palermo and Messina in prodigious quantities. There is no striking difference between this and *A. Tobianus,* excepting the extraordinary shape of the dorsal fin, which is invariably undulated and narrowed in the middle. It never grows to a size exceeding the figure, and is usually much less; while the British species is often found double the length. Like that, also, *A. Siculus* has the lateral line running close to the dorsal fin; for the fine line in the middle of the side, as Lacepede has well observed, is that only which connects the muscles. That author likewise mentions, that the jaws in *A. Tobianus* have minute teeth, but these I could never discover. The rays of the fins are, pect. 16; dorsal 56; anal 30.

Common Sand-Lance—right figure.

Dorsal fin linear, equal. [*Linn. Syst. Nat. vol.* i. *p.* 1145. *Pennant* iv. *pl.* 28. *Bloch, pl.* 73. 2.] [*Lacepede,* ii. 275. *Klein Hist. Pisc. fasc.* iv. *tab.* 12. *f.* 10.]

This, though a very common fish, has been figured by all authors as if the rays were spined and naked at their extremity; they are, on the contrary, soft and connected.

It abounds at certain times on many parts of our coasts. The number of rays stand thus: Pectoral 12; dorsal 51; anal 27.

64

64. *Macroglossum assimile.*

Generic Character.

Antennæ subfusiform, gradually thickest towards the end, the tip abruptly terminating in a very short slender incurved hook; ciliated in the male, simple and more slender in the female. Palpi porrected, thick, the last joint pointed. Wings opaque, entire. Abdomen thick, bearded. Generic Type *Sphinx stellatarum.* Linn.

Specific Character.

Wings brown; anterior with two darker obscure bands, and a terminal blackish dot; middle of the posterior wings, and three lateral spots on the body, orange.

This genus was instituted by Scopoli many years ago, and differs principally from *Sesia* by having opaque wings, and from *Thyreus* by the very great difference in the construction of their antennæ. Many exotic species are known, but only one is found in Europe (*Sphinx stellatarum* of Linn.), which likewise inhabits our own country, and to which this our insect is very nearly allied.

Though by no means uncommon in collections, I cannot find this species either figured or described, nor indeed am acquainted with its locality. The under figure is of the male; the upper of the female; which differs only in the wings being rather broader, and in having one segment in the body less than in the other sex.

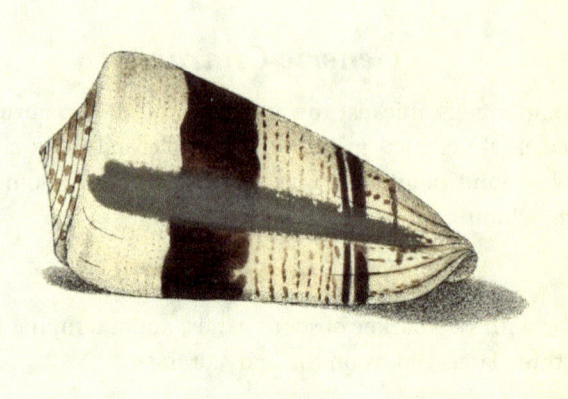

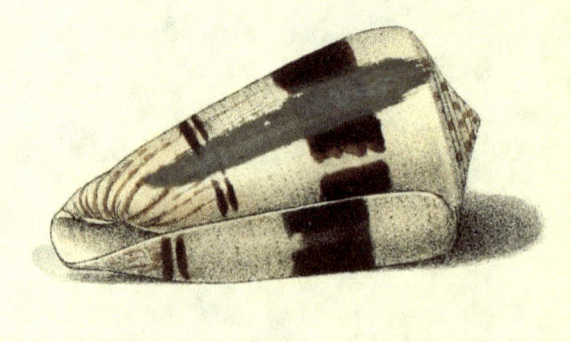

65. *Conus Augur.*
Girdled Cone.

Generic Character.

Shell univalve, turbinated, inversely conic, convolute; aperture longitudinal, narrow, not toothed; base effuse; spire very short. Generic Type *Conus marmoratus*, &c.

Specific Character.

Shell smooth, fulvous-white, with broad dark chesnut bands and transverse lines of dots; spire ob- tuse, convex, striated, depressed. [C. augur. Lamarck Annal. Mus. xv. 277. Encycl. Méth. 333. 6.] [Conus magus. Gmelin 3392. 57. Martini ii. 58. 641.]

The Girdled Cone is conspicuous among the beautiful shells of this extensive family, by the broad and rich chesnut bands, which are either two or three in number, and more or less broken into spots; in high-coloured shells the minute lines of dots between them are also of the same colour. It is not a common species, and inhabits the Asiatic ocean.

This is the *Conus Magus* of Gmelin and Martini, and of our sale catalogues: this error has orig- inated from Gmelin having described two distinct shells, *C. Augur* and *C. Magus*, under the latter name.

I have not referred to Lister's figure 755. 7, being doubtful of its affinity; and those of Martini and Bruguieres are very bad.

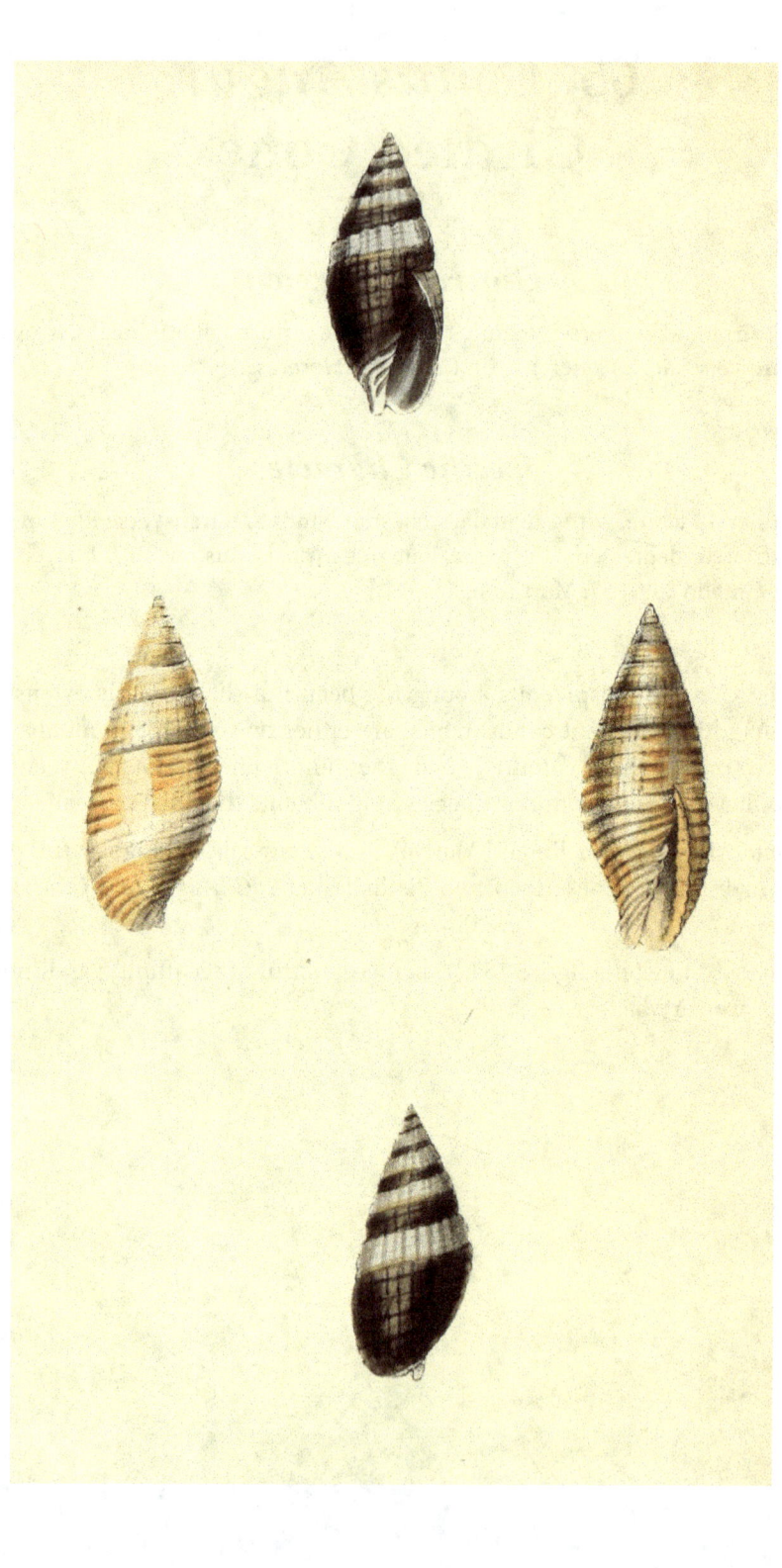

66. *Mitra lugubris.*
White-banded Mitre.

Generic Character.

See Plate 23.

Specific Character.

Shell inversely pear-shaped, brown, with transverse sulcated grooves, punctured within; volutions above obsoletely plaited and banded with white; outer-lip thin, margin crenated; pillar four-plaited; base white, truncated.

We have had much difficulty in the investigation of this species: for its characters cannot be reconciled with any of those contained in Lamarck's Monograph of the genus in the *Annales du Museum*. With regard to the unnamed figures in the old authors, it bears the closest resemblance to that of *Gualtieri*, tab. 32. G, which Lamarck quotes for his *M. crocata*; but then his description is not at all applicable to our shell; and Mr. Dillwyn's synonyms of the Linnean *V. nodulosa* (where he has also included *M. crocata*), we are satisfied comprises two or even three distinct shells.

This was named by Dr. Solander from the specimen in Mr. G. Humphrys's collection here figured: it is exceedingly rare, and its locality unknown. In form it resembles a *Buccinum*; the transverse grooves are broad, strongly defined, and have large and deep excavated dots within them; the upper part of each whorl has an appearance of irregular plaits, which makes the suture uneven, and takes off something from the smoothness of the lower part of the whorls, but the shell is in no way granulated.

MITRA ferruginea. Thick-lipped Mitre.

Specific Character.

Shell clouded and spotted with ferrugineous, with transverse elevated ribs; outer lip thick, obtusely crenated; pillar four-plaited. [M. ferruginea. *Lam. Ann. du Mus. vol.* 17. *p.* 200.] [*Young.* Vol. vitulina. *Dill.* 553.—*Martini* 4. 149. 1380 & 1.] [*Variety* more elongated. Vol. abbatis. *Dill.* 557. *Chemnitz* 11. *t.* 177. 1709 & 10.]

This (a common shell) can be no other than the *M. ferruginea* of Lamarck, though neither that author nor any other has noticed its primary distinguishing character, that of the uncommon thickness of the outer lip at the margin, which is also divided into convex obtuse crenations; in young shells this is not apparent; such is Martini's figure. Mr. Dillwyn has changed Lamarck's name to *Vitulina* for this, and given the name of *Abbatis* to the variety more lengthened, figured by Chemnitz; but which, from specimens now before us, we consider with Lamarck only as a variety, possessing all the essential characters here given to both.

Editor's note

The volumes of William Swainson's *Zoological Illustrations* are testimony of a glorious era of explorations and scientific discoveries at the beginning of the XIX century.

European naturalists traveled exotic places to bring back specimen and description of animals and plants never seen before.

A great effort was spent to categorize them in a precise taxonomy that would show the differences and similarities of those new species. A competition, really, to win the honor of being the first to name a new bird or a new plant.

The rich illustrations in this volumes, the detailed descriptions and the author's comments open a window on that time.

This new edition is not just a reprint of the original work, it is meant to make it easier, for the contemporary reader, to enjoy the vivid images and immerse in the spirit of a now forgotten time.

Either by just flipping through the images or by carefully reading the descriptions, we hope you will enjoy this book as we enjoyed giving them a new life.

January, 2021

www.ingramcontent.com/pod-product-compliance
Lightning Source LLC
Chambersburg PA
CBHW081512220526
45467CB00010B/2888